Penguin Handbooks

THE BACK GARDEN WILDLIFE SANCTUARY BOOK

Ron Wilson is Head of Everdon Field Centre in Northamptonshire, and Chairman of the Northamptonshire Association for
Environmental Education. He is Honorary Education Officer of
the Humane Education Centre, and Editor of the NAFSO
Journal. As well as having a keen practical interest in the protec-
tion of the environment, he is also the author of several books
on wildlife and related topics, including *The Hedgerow Book*,
Wild Flowers, *Vanishing Species*, *A Year in the Countryside*, *The
Nature Detective's Notebook*, *The Life of Plants* and *An Urban
Dweller's Wildlife Companion*. He is also working on a series of
Young Naturalists Books. The American edition of *How the Body
Works* has been awarded a certificate as an Outstanding Science
Book for children by the National Science Teachers' Association
of America. Ron Wilson also writes a regular feature, 'Country
Scene', for the *Daventry Weekly Express*, and has contributed
to a number of magazines and journals.

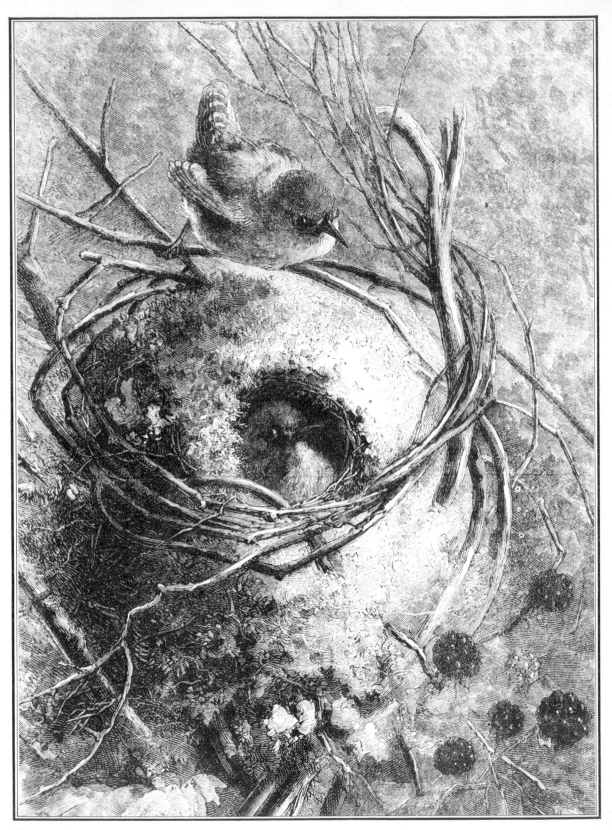

The wrens' nest

THE
BACK GARDEN
WILDLIFE
SANCTUARY BOOK

RON WILSON

with new line drawings by Anne Roper
and John Heritage

PENGUIN BOOKS

Penguin Books Ltd, Harmondsworth, Middlesex, England
Penguin Books, 625 Madison Avenue, New York, New York 10022, U.S.A.
Penguin Books Australia Ltd, Ringwood, Victoria, Australia
Penguin Books Canada Ltd, 2801 John Street, Markham, Ontario, Canada L3R 1B4
Penguin Books (N.Z.) Ltd, 182–190 Wairau Road, Auckland 10, New Zealand

First published by Astragal Books 1979
Published in Penguin Books 1981

Made and printed in Great Britain by
Butler & Tanner Ltd, Frome and London
Set in Monophoto Plantin

CONTENTS

ACKNOWLEDGEMENTS

This book would not have been possible without the help and encouragement of a number of people. The idea stemmed from some articles which I did for Bernard Schofield and Andy Pittaway's *Country Bizarre's Country Bazaar* and *The Complete Country Bizarre*. For the opportunity to use the pages of their magazine and later their books I am grateful. To Anne Roper and John Heritage who have spent endless hours slaving over a hot pen to produce the illustrations; to all those organizations which have answered my letters/queries, *etc*. – for their help and advice I am grateful, and for the interest which they have shown in the project. To the Henry Doubleday Research Association for permission to reproduce their Hedgehog House on page 55 and the Royal Society for the Protection of Birds for permission to reproduce the log nest-box on page 39, also to the Society for the Promotion of Nature Conservation for the bat box on page 59. To Lyn Reynolds for typing the manuscript, to Godfrey Golzen for the loan of many ancient books, and finally to Alexandra Artley formerly at Astragal Books and Margaret Crowther (editor) for their enthusiasm and help at all times.

INTRODUCTION

You cannot have failed to notice, whether you are a dedicated naturalist or simply a lover of nature, that there is less of our countryside for wildlife. Many hedges are gone, and the intricate web of wildlife which they harboured is no more; deadly pesticides ensure that once common weeds will no longer flourish, and many harmless invertebrates alas have died alongside the harmful pests.

Very soon, perhaps sooner than we dare to imagine, parts of our island – once rich in wildlife – will no longer harbour those common plants which we have come to accept as part of our heritage. Many knowledgeable naturalists, as well as general nature lovers, have grave fears for the very survival of even the most common species in Britain, let alone the rarer ones.

Technological advances and modern society go hand in hand, and both decree to a large extent what sort of life we lead, and where we live. Modern planners, and no one seems to understand their reasoning, place houses far back from the road, with a large expanse of nothingness at the front. Most have to be set with lawns, and any flower beds are small. Local dogs and even children seem to claim these spaces as communal property. And behind the house? A small patch, often in a terrible state due to the builders' rubble.

What we hope to do in this book is to suggest ways in which people can turn part or all of their gardens into a wildlife sanctuary: a place which will, for example, support nettles, encourage insects and offer a 'supplementary habitat' for some of our threatened wildlife species. We hope that with the present vast destruction of large tracts of our countryside, *you* will consider doing *your* bit to offer a 'place for wildlife'.

In such a book as this there is a lot which we have had to leave out. We have tried to remedy this to some extent by summarizing certain aspects, like the types of trees which you could grow, the species useful for hedges, the birds which *might* visit the garden, and so on.

We are also aware that all the things which are suggested are not practicable in all situations. If you have an alternative, please let us know, so that, if the book is reprinted, we might be able to include it.

Most of the items which you need can be obtained from your local garden centre, and you will find details of these in the *Yellow Pages* of your local telephone directory. On the other hand, you might live in an isolated place and prefer to send off for some items: others are only obtainable by post. This information, together with details of suitable books, *etc*, is listed at the end of each chapter under the heading *Further Useful Information*. We hope that the inclusion of this material will be helpful, and that it will assist with the planning of your own distinctive wildlife garden.

We hope that you will enjoy this book, and that you will be able to do your bit for wildlife, however small and insignificant a part you might think you can play. If you want to write to us about your efforts, offer comments and criticisms, please do so – you can contact me c/o Astragal Books, 9 Queen Anne's Gate, London SW1H 9BY. I look forward to hearing from you.

Weedon, 1979   Ron Wilson

THE NATURAL WILDLIFE GARDEN

Perhaps it would be worthwhile pausing and posing the question, 'How did our garden grow?' To consider our garden a wild patch we may have to go back several hundreds of years. Slowly, over the centuries, as man has become civilized he has decreased the rural wilderness and turned it into an urban brick and concrete jungle. You will probably not know what was on the site of your house and garden. But it might have been a heath; it could have been marshland, or perhaps it was even a wood. The place would almost certainly have been teeming with wildlife in its heyday, but where has this wildlife gone? Many species have become extinct, and are lost for ever as part of the traditional pattern of the intricately woven web of wildlife which abounded when Britain was a wild island. More than ever, we need to plan our garden to support wildlife and that's where this book comes in . . .

A Poison-free Environment

It is not so very long ago that the major tools for keeping down the weeds in the garden were the hoe and spade. Very hard work, when compared with today's modern press-button spray. As someone who cares about wildlife and particularly about the garden as a wildlife sanctuary, your first principle should be that the hoe is mightier than the spray. To preserve the wildlife already in our gardens and to offer a sanctuary for those creatures which might want to take up residence, we need to maintain our garden by natural means. Keeping a natural garden is not as difficult as it may seem, and your garden could be as neat and tidy as your neighbour's, sprayed with every conceivable killer. The only thing you need to do is to convert your neighbour to your way of thinking, so that his poisonous sprays do not drift into your garden and attack its inhabitants. You could perhaps start a simple revolution in your neighbourhood, and have a poison-free gardening environment.

Have you ever considered what would happen in the garden if you did nothing? You would still have a crop, because everything would continue to grow, but your crop would probably not be the sort which you expected. You would have a garden full of what most people would call weeds, but which the nature lover would probably see as valuable food for a whole array of wildlife. And what is a weed? The answer is quite simply 'a plant which grows where man does not want it'. In other words it is the right plant growing in the wrong place. But what is the right place you might ask? Almost certainly if we found these weeds growing in the hedgerows we would call them wild flowers.

If we take the 'wrong place' idea to its logical conclusion, you will soon see that most of the plants in your garden shouldn't be there. Many of the cultivated species which do so well are not native to our shores, but to lands far distant from these islands. So really even your most treasured plants are probably weeds, because they are flowering in the wrong place! You don't really have to feed the natives like sow thistle and dandelion with fertilizers; they will grow only too well without these additives!

Even if you live in a new house, where the garden is also recent, the chances are that the soil has been growing crops of one sort or another for centuries. Yet still, and very faithfully, the soil continues to do its best to support the vast number of plants which we manage to put there. There are four 'free' aspects to the garden: the soil, water, air and sun. All are important for the well-being, not only of the soil, but of the life which it supports – and we must remember that as well as having the plants which in most cases we can readily see, there are hundreds of animals living below the surface, in the soil, out of our sight.

The Soil

The productivity of your garden will depend on the soil and having knowledge of the make-up of your soil will stand you in good stead when it comes to working out the sort of plants which you can grow for wildlife to enjoy. If you are lucky and your house has been built on rich loam you should be laughing; if you are on clay, you might not be quite so happy! The type of soil which you have is due to its structure, the way in which the particles are made up: and that was determined many centuries ago, when the rocks were laid down. For it is to the rocks that the soil owes its origins.

But your soil wouldn't be soil, whether it is good

or bad, were it not for the sterling work going on unseen below its surface. Here the earthworm continues on its unending, albeit unwitting, job of acting as nature's ploughman. But worms aren't the only beings important to your soil. There is a whole host of other 'creepy-crawlies' helping too. If you turn over the soil you will probably expose a few of them, although with an air of urgency they scuttle off, to seek refuge once more in the inky blackness. Those big enough to see we know about, but there are many more whose activities we will never witness, although we will see the result of their labours. The bacteria and the fungi, the former microscopic, and many of the latter also, are breaking down the materials which have fallen onto the soil, so that fertility will reign supreme. The cycle of events is very important and this cannot be over-emphasized. The plants which took their nutrients from the soil last spring will complete their own cycle. Once it is over their leaves will fall, and the activities of the myriad soil creatures will ensure that decay is brought about and the materials returned to the soil, ready to be used again.

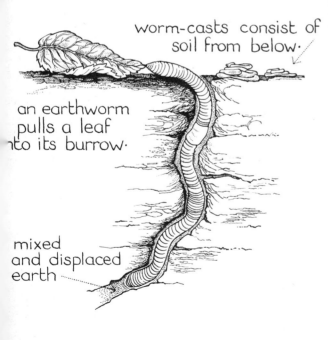

worm-casts consist of soil from below.

an earthworm pulls a leaf into its burrow.

mixed and displaced earth

The work of the earthworm

Compost and Composting

For centuries man has used the soil, from which nutrients have been taken as he has grown his crops. We often forget that in the natural state of things life and death form a complete cycle. The events which go on in our gardens are also part of this cycle of activities. In simple terms it works in this way: as the plants grow they take nutrients from the soil, and in the complicated process of photosynthesis they manufacture – synthesize – food, using sunlight and carbon dioxide. Animals eat the plants as a source of food, some animals will eat other animals, and some eat both plants and animals. Inevitably both plants and animals die, and in due course decay. In this way the nutrients will be returned to the soil, so that they can be used by other plants. But often this cycle is not completed 'naturally', as happens in the wild, in a wood for example. The reason for this is that the gardener takes up his crops. The nutrients used to make food, and stored, for example, in the potato, means less material in the soil. In recent years the easiest way to replace the nutrients has been by the addition of artificial fertilizers. Too often the natural process has been eliminated and the compost heap has been forgotten. We are too keen to throw everything into the dustbin. We clear up the garden in autumn, burn the leaves, and tip the clippings into the hedge.

A free supply of plant food
Yet here is the natural means of obtaining a regular and free supply of compost which is so valuable in our gardens. Some people envisage the compost heap as an untidy pile of rubbish which has to be pushed out of sight into the corner of the garden, to prevent it from becoming an eyesore. But there is no need to use this old-fashioned method. Modern materials have made it possible to produce compost bins, which are relatively cheap to purchase. As an alternative it is possible for the gardener to make a simple one. A number of firms sell compost bins direct to the public, and we give the names of some manufacturers at the end of this section so that you can write for details. The compost bin is scientifically designed so that the rubbish which you throw in will come out a few months later as a very nutritious compost or manure, which will do your plants the world of good. The ideal number of bins is three. With three there will be one which is full,

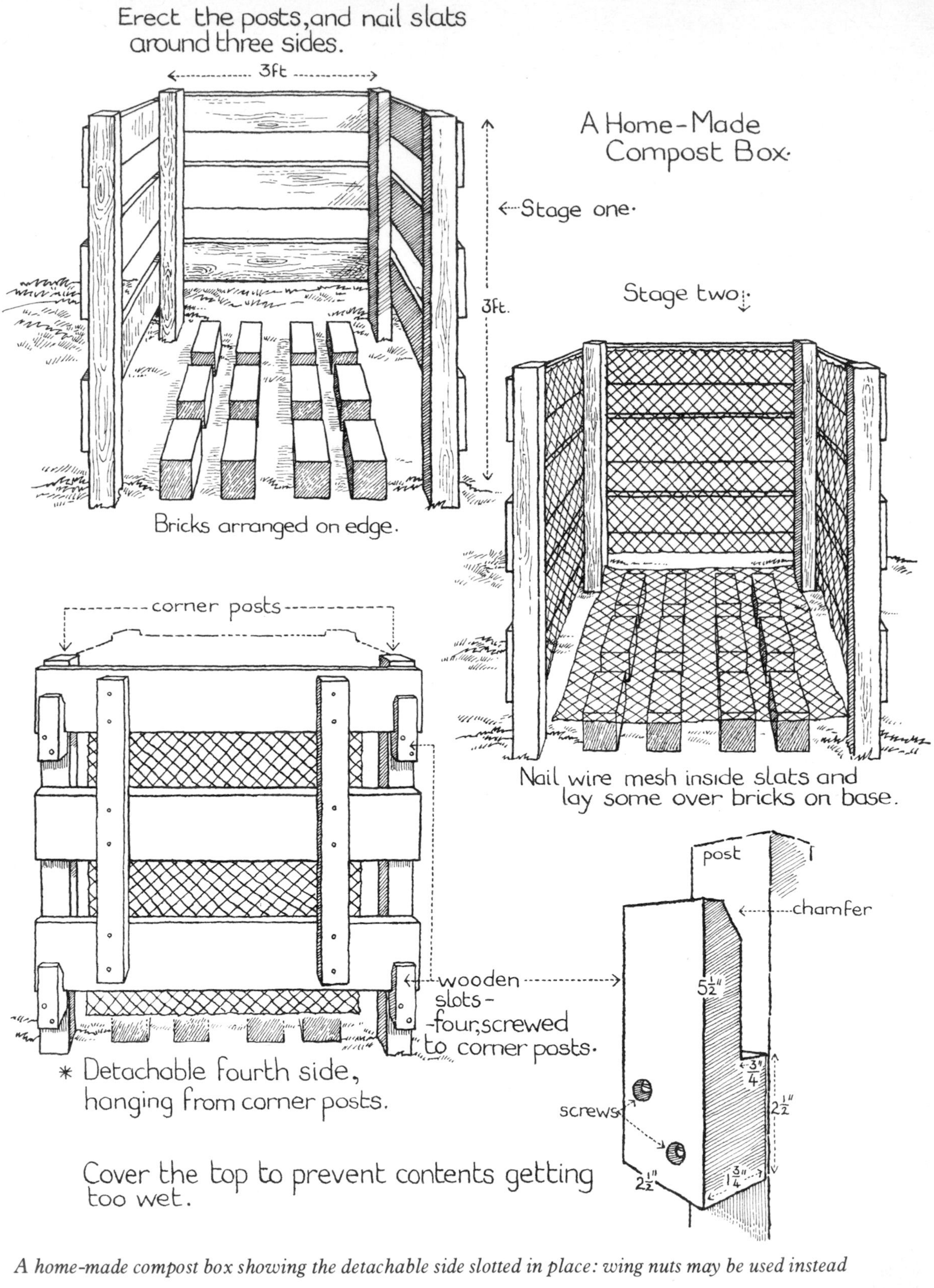

Erect the posts, and nail slats
around three sides.

3ft

A Home-Made
Compost Box.

←— Stage one.

3ft.

Stage two.↓

Bricks arranged on edge.

corner posts

Nail wire mesh inside slats and
lay some over bricks on base.

post

chamfer

$5\frac{1}{2}''$

$\frac{3}{4}$

$2\frac{1}{2}''$

screws

wooden
slots -
-four, screwed
to corner posts.

$2\frac{1}{2}''$

$1\frac{3}{4}''$

* Detachable fourth side,
hanging from corner posts.

Cover the top to prevent contents getting
too wet.

A home-made compost box showing the detachable side slotted in place: wing nuts may be used instead

4

with the compost in use or ready to be used. The second will probably be about half full, and the third one will be empty, ready for another heap when the second one is maturing.

Build your own bin

If you are going to have more than one bin you will probably decide to build your own, since this is by far the cheapest way. The most important criterion to remember is that the bin must *not* be solid. You must have plenty of air circulating. In fact for a bin to be at its most efficient there should be as many holes as solid material. The best idea is to have a look around garden centres so that you can see what models are available. Barrel-shaped bins are the most suitable; there are no corners and the air can circulate easily and evenly. Old oil drums *thoroughly* cleaned are often cheap to acquire, and all you have to do here is to ensure that you have plenty of holes. If you are going to make your own bins from scratch, you will probably find it difficult to construct a barrel-shaped container. Although they are slightly less efficient, square bins can be used. It is tempting to join them together for convenience, but if you do this you will stop air from getting at your material on at least one, and possibly two sides and your material will not compost properly.

Materials to use

Having decided to build your own bin, what do you make it from? There is a wide range of materials to choose from. Breeze blocks, stone, concrete slabs, brick, wood, rigid PVC sheets, all will serve the same purpose. Whatever material you decide to use you should remember that air spaces are vital. This is especially important as far as the bottom of the bin is concerned. Place the container on breeze blocks, bricks or planks, so that air can circulate below it. You will also need to be able to remove at least part of one side of the bin, otherwise you won't be able to get at your compost – you will also find it advantageous when turning your material. If using wood the simplest way of fastening the removable side is by using wing nuts. With breeze blocks or bricks all you will need to do is remove one side. Once your bin has taken shape you will need to cut chicken wire to size, so that it will fit inside the bin. This will prevent the contents from falling out through the gaps. The plastic coated variety is best, as rusting will not be a problem.

Filling the bin

What sort of materials can you throw into your compost bin? If we refer back to our woodland example you will soon see that over a period of time *everything* in the wood will eventually die and decay and all the animals in the wood, including of course the birds, will regularly add their droppings. If you walk into most woods you will usually find a good layer of leaf mould, where the plant element is decaying. So in the garden there is virtually no limit to what you can throw into your compost bin. Naturally you will add all the garden refuse, from lawn clippings to rotting vegetables. Avoid putting in diseased plants; these should be burnt to ensure that the disease is destroyed. There are one or two other plant items which should be excluded: soft plant tissue, like that which makes up the leaves of grass and trees, decays fairly rapidly, but this is not true of the stems pruned from the hedge or from the rose bush. It has been estimated that the period needed for their decay can be up to eight times that of the leaves. You can burn these and save the ash, because this can be used in the garden, providing a valuable source of nutrients.

Apart from the wide range of material which comes from the garden the chances are that there will be almost as much waste from the kitchen. The potato peelings, tea leaves and pods from beans and peas all make valuable contributions. Cardboard boxes can be torn up and placed in the bin, and don't forget that the excrement from your pets or other livestock should be added too. If you have seen an animal carcase you will realise that the harder parts, like the bones, take a long time to decay; much longer than the plant material. This is one reason why other forms of animal material should be avoided; it is fairly safe to assume that it will not have rotted down by the time the rest of your compost is ready. There are other reasons as well. The smell given off will attract animals from round about. Although you might get a fox as well as local cats and dogs, you do not really want them scavenging in your compost bin, and upsetting the natural development.

The process of decay

Perhaps it is a good idea to understand briefly what happens as the compost is made from the 'waste' materials. Bacteria are responsible for bringing about the decay, together with some fungi. There are many different types of bacteria, and if the decay process is to work well then you need more than one kind. Variety of material is the keynote of successful composting. Make sure that you place your compost bin in the sunlight, so that maximum activity will take place. During long periods of dry weather the compost heap will dry out. The result is that the bacteria will either die or stay in the inactive state, so make sure that your compost is always damp, so that its activities can progress unhindered. The other extreme also has an undesirable effect. Should the material in the bin get too wet, the bacteria will not work efficiently, if at all, so make sure that the top of the bin is covered in very wet conditions. You can use a piece of wood or plastic sheeting. A few bricks will ensure that the cover isn't lifted off in high winds.

Speeding the activity
There are many different types of commercial activator on the market, and their function is to speed up the composting process. But there really isn't any need to buy them, when there is a free supply! If you live near a farmyard, you will probably have access to farmyard manure. If you can obtain some this helps to speed up the reaction in your bin. No farmyard nearby? No need to despair if you have some small mammals or birds as pets. Their droppings serve the same purpose. You haven't any of these either? Never mind, you still have one course of action, and that is to use human urine – and there's plenty available, although you might have difficulty in collecting it!

A year-long process
From spring through to autumn the bacteria will be working away happily in your bin, but they, like many other forms of life, don't take too kindly to the drop in temperature winter brings. In fact if bacterial activity doesn't actually stop in the colder months of the year, it will be greatly reduced. To ensure that the process will continue during the winter you will need to keep your bin warm by surrounding it with heavy-duty black polythene.

Using your compost
Once the compost is ready – and you can tell by its dark brown colour and crumbly nature – you will be able to use it. Place it on the top of the soil and those 'natural ploughmen', the worms, will carry out their duties, dragging the material below the surface of the soil. Where the earthworm has been working you will often see little piles of soil – the worm casts – and they are particularly visible on the lawn. As the worm has been working below the soil, it has tunnelled, and mixed up the soil in doing so. You don't actually need to exert yourself by digging in the compost if you place it around your plants – the hard work will be done for you by the obliging worms.

Your Wildlife Garden And The Law

A knowledge of the law concerning wild plants and creatures is particularly important when we are considering having wild plants in our garden, especially if we intend to transplant some. For many years now there have been by-laws relating to wild plant species. Many have decreased dramatically: others have vanished altogether. Ultimately, it was not surprising that we should need a Conservation of Wild Creatures and Wild Plants Act. Coming onto the statute books in 1975, this Act gives some form of protection to almost every plant species in the country, except for the more persistent (noxious) weeds, which many naturalists would prefer to term wild plants. So what is the situation? If you want to dig up *any* wild plant you must have the permission of either the owner or the occupier. Wild plants in this instance include flowering species, mosses, liverworts, lichens, ferns, but not mushrooms and toadstools. This might seem rather drastic, but you are of course allowed to weed your garden, if you want to do so. And you can go into the countryside in order to gather the wild fruits and berries, like elderberries and blackberries. But there are twenty-one plants which are now so rare that so much as to touch them is an offence.

Please remember that it is now an offence to uproot any wild plant, without good reason.

For further information you can consult the leaflet which has been produced by the Council for Nature, and called *Wild Plants and the Law*. It is available free of charge (a stamped addressed envelope would be appreciated) from Council for Nature, c/o Zoological Gardens, Regent's Park, London NW1 4RY.

An alternative leaflet, giving more detailed

information, is available, price 20p, from British Museum (Natural History), Cromwell Road, London SW7 5BD, entitled *Wildlife, The Law and You*. For further information on wild birds and the law see page 12 of this book.

Further Useful Information

Organizations and Societies

Bio-Dynamic Agricultural Association, Broome Farm, Clent, Stourbridge, West Midlands, DY9 0HD

The primary object of the Society is to foster and promote the agricultural impulse given by Rudolf Steiner in connection with bio-dynamic agriculture. Further details of membership (sae) from the above address. Publishes *Star & Furrow*.

Botanical Society of Edinburgh, c/o The Royal Botanic Garden, Edinburgh 3

Scotland's national botanical society. Details of membership available.

Botanical Society of the British Isles, Hon. General Secretary, c/o Department of Botany, British Museum (Natural History), Cromwell Road, London SW7 5BD

Concerned with the study of British flowering plants and ferns. The organization also publishes a number of periodicals and books, including *Watsonia*, *BSBI Abstracts* and *BSBI News*. A book advising on the native trees and shrubs for naturalization, *Plant Native Trees and Shrubs*, is to be published later in 1979. Further details of membership, *etc*, from the above address.

British Naturalists Association, Hon. Membership Secretary, 23 Oak Hill Close, Woodford Green, Essex

The organization is made up of large numbers of interested people scattered throughout the length and breadth of the British Isles. With a wide interest in all aspects of natural history, they are able to contribute to the society's journal. There are groups in various parts of the country. The invaluable series of leaflets, *How to Begin the Study of* . . . is published by BNA.

Council for Nature, Zoological Gardens, Regent's Park, London NW1 4RY

This organization attempts to bring together the interests of amateur natural history societies throughout the country, and individuals may join to aid with the work. Members and associates receive the newsheet *Habitat* at regular intervals.

Council for the Protection of Rural England (CPRE), The Secretary, 4 Hobart Place, London SW1Y 0HY

The aim is to organize concerted action to 'improve, protect and preserve the rural scenery and amenities of the English countryside, its towns and villages, and to act as a centre for furnishing of advice and information'. Further information from the above address.

Garden History Society, Hon. Secretary, 12 Charlbury Road, Oxford, OX2 6UT

The Society was formed to bring together those who are interested in the various aspects of garden history – horticulture, landscape design, forestry and related subjects. It publishes *Garden History*, which is an illustrated journal, and holds seminars, arranges tours, meetings, *etc*.

Good Gardeners Association, The International Association of Organic Gardeners, General Secretary, Arkley Manor, Arkley, Barnet, Herts, EN5 3HS

When a gardener joins he will be taught how to 'garden nature's way'. The organization runs experimental and demonstration gardens at Arkley Manor. Members are allowed to visit these and to receive other benefits, including bi-monthly bulletins and newsletters. Lectures are arranged and advice and free soil analysis are also part of the rights of becoming a fellow.

Society for the Promotion of Nature Conservation (SPNC), The Green, Nettleham, Lincoln, LN2 2NB

The major voluntary organization in Britain concerned with all aspects of nature conservation. Forty of the Counties' Nature Conservation Trusts are affiliated to it. Individual membership is available. Publishes *Conservation News*.

Soil Association, Walnut Tree Manor, Haughley, Stowmarket, Suffolk, IP14 3RS

A world-wide charity formed in 1946 to promote a fuller understanding of the vital relationship between soil, plants, animals and man. The Association believes that these are parts of one whole and that nutrition derived from a balanced living soil is the greatest single contribution to health (wholeness). For this reason it encourages an ecological approach and offers organic husbandry as a viable alternative to modern intensive methods.

Details of membership freely available.
Wildflower Society. Mrs V. V. C. Schwerdt, Rams Hill House, Horsmonden, Tonbridge, Kent

Concerned with all aspects of the study of wild flowers.

Periodicals and Magazines

All Living Things, Humane Education Centre, Avenue Lodge, Bounds Green Road, London N22 4EU

This is the magazine for younger members of the Crusade Against All Cruelty to Animals Ltd.

Country-Side, Hon. Membership Secretary, British Naturalists Association, 23 Oak Hill Close, Woodford Green, Essex

Published three times a year, it is available free to members, but can be purchased by other interested people.

Habitat, Council for Nature, Zoological Gardens, London NW1 4RY

Published nearly every month, and sent to Associates and Members. It keeps everyone up to date with 'nature' affairs.

The Living World, Humane Education Centre, as above

This is the adult Journal of the Crusade Against All Cruelty to Animals Ltd, and contains informative articles about the aims of the society, as well as general natural history material.

Oasis, P.O. Box 237, London SE13 5QU

The magazine of conservation gardening, this publishes important articles about planning and planting, as well as general informative material about the plants and animals which might occur. It is published (on subscription) six times a year.

Watchword, The Watch Trust for Environmental Education, 2 The Green, Nettleham, Lincoln

This is aimed at young people who are interested in their environment.

Wildlife Magazine, 243 Kings Road, London SW3 5EA

Published monthly, and available from newsagents or by subscription, the magazine contains many useful articles about natural history in this country, as well as abroad. It is also the 'organ' for the Wildlife Youth Service, and contains special sections each alternate month.

Books, Leaflets, etc

AA Book of the British Countryside, Drive Publications/Readers Digest

Baker, M., *The Gardener's Folklore*, David & Charles

Binding, G. J., *Organic Gardening and Farming*, Thorsons

Bruce, M. E., *Commonsense Compost Making*, Faber

Campbell, S., *Let it Rot! The Home Gardener's Guide to Composting*, Thorsons

Chinery, M., *The Family Naturalist*, Macdonald & Jane's

Complete Book of the Garden, Readers Digest

Conservation of Wild Creatures and Wild Plants Act, 1975 This information leaflet is available from the Council for Nature, address earlier in this section. Please send a stamped addressed foolscap envelope

Darwin, Charles, *On Humus and the Earthworm*, Faber

Dennis, E. (Ed.), *Everyman's Nature Reserve*, David & Charles

Ellis, E.A., *Wild Flowers of the Hedgerows*, Jarrold

Ellis, E. A., *Wild Flowers of the Field and Garden*, Jarrold

Fertility Without Fertilizers, Henry Doubleday Research Association. See address earlier in this section

Friedman, J., *Complete Urban Farmer*, Fontana

The Gardening Year, Readers Digest

Genders, R. *Scented Wildflowers of Britain*, Collins

Genders, R., *Wildlife in the Garden*, Faber

Grigson, Geoffrey, *The Englishman's Flora*, Paladin

How to Begin the Study of Series, British Naturalists Association. Address earlier in this section

Keeble Martin, W., *Concise British Flora in Colour*, Ebury Press/Michael Joseph

Mabey, R., *The Unofficial Countryside*, Collins

Margoschis, R., *Recording Natural History Sounds*, from Print & Press Services, 69 Beech Hill, Barnet, Herts, EN4 0JW

Nicholson, B. E., Ary, S. & Gregory, M., *Oxford Book of Wild Flowers*, Oxford University Press

Owen, Denis, *The Natural History of Britain and North Europe — Towns and Gardens*, Hodder & Stoughton

Pest Control Without Poisons, Henry Doubleday Research Association

Prime, C., *Plant Life* (Countryside Books), Collins

Rodale, J. I., *The Complete Book of Composting*, Rodale Press

Russell, Sir J., *The World of the Soil*, Collins

Salisbury, Sir J., *Weeds and Aliens*, Collins

Schofield, B. & Pittaway, A., *The Complete Country Bizarre*, Astragal Books
Schofield, B. & Pittaway, A., *Country Bizarre's Country Bazaar*, The Architectural Press
Shewell-Cooper, W. E., *Soil, Humus and Health*, David & Charles
Shewell-Cooper, W. E., *Compost Gardening*, David & Charles
Soper, T., *Wildlife Begins at Home*, David & Charles/Pan
Stokoe, W. J., *The Observer's Book of Wild Flowers*, Warne
Wilson, Ron, *A Year in the Countryside*, Artworks

Equipment and Supplies

Ardenco Ltd, Roseberry Avenue, Melton Mowbray, Leics:
   Compost maker
Bees Ltd, Sealand, Chester:
   Bees Composa – Compost maker
Chase Compost Seeds, Benhall, Saxmunden, Suffolk:
   Compost grown flower and vegetable seeds
Compton Buildings Ltd, Fenny Compton, Leamington Spa, Warwicks:
   Concrete compost bin
Dobie, Samuel & Sons Ltd, Grosvenor Street, Chester:
   Untreated seeds
Hawkins, A. R., 2 The Parade, Northampton:
   Binoculars at discount
Heron Optical Co, 23-25 Kings Road, Brentwood, Essex, CM14 4ER.
   Binoculars and telescopes
Humex Ltd, 5 High Road, Byfleet, Surrey, KT14 7QF:
   Miniature propagator

Hunter, James, Ltd, Chester:
   Bulk supplies of flower and vegetable seeds – compost grown
Hurst, Gunson, Cooper & Taber Ltd, Coggeshall, Essex:
   Untreated seeds
King, E. W. & Co. Ltd, Colchester, Essex:
   Untreated seeds
Metwood Accessories, Broadacre, Little Linford Road, Haversham, Wolverton, Bucks:
   Telescopes and binoculars
Rotocrop Ltd, 120 Worthing Road, Basingstoke, Hants:
   Rotocrop Compost bins
Rutland Products, Manor Lane, Langham, Oakham, Leics:
   Compost maker
Sudbury, Corwen, Clwyd:
   Soil testing kits

Miscellaneous

Conservation Books, 228 London Road, Reading, Berks, RG6 1AH
Has a wide range of conservation titles, but can also obtain any other book.
Country Book Club, Readers Union Ltd., P.O. Box 6, Newton Abbot, Devon, TQ12 2DW
Members are able to buy many country/conservation/natural history books at discount. There are also special offers.
Country Book Society, address as for previous entry
Offers books by original publishers at discount to members.
World of Nature (Book Club), P.O. Box 19, Swindon, Wilts, SN1 5AX
This quarterly book club offers readers hardback books at discounted prices.

BIRDS
IN THE
WILDLIFE
GARDEN

The illustration shows (clockwise, from top):
flying blackbird, elderberries, small rookery in trees,
magpie, house martin's nest under house eaves,
bird bath on lawn, song thrush using wall as an
anvil to crack snail shells, cotoneaster berries, blue
tits on a string of peanuts

In assessing the value of the garden as a valuable nature reserve there are some interesting statistics which will prove its value. Observations have shown that the blackbird, for example, prefers the confines of the garden to the open countryside. For every blackbird found on farmland, there may be twenty found in gardens. You will probably be surprised to learn that even in gardens in our larger cities some keen naturalists have counted as many as twenty-one different species of birds coming to feed and nest.

To understand this situation we have to remember that most of our natural countryside was once woodland. Slowly over the centuries the vast tracts of tree-covered countryside have gone and the birds have slowly adapted to the changing situation. Instead of living in afforested areas they took to small groups of trees and then to solitary tree species. So it is not surprising that the garden, often with trees in its hedges, and with berry-bearing shrubs which are valuable for food, is a man-made habitat which birds have found valuable. The chaffinch, dunnock, wren and tit, which we often label garden birds, once dwelt in our woodlands: now they have a new home. And if we can make our gardens conducive to their requirements, so much the better, and they are likely to become regular visitors and even residential when they will breed.

Naturally the situation of a garden, not only in terms of its position in an area, for example on the edge of a town, or in the centre of an urban conurbation, but its geographical site within the country will determine to some extent which birds make use of it. Surveys have shown that there are now around fifty different species which will inhabit our gardens. While there are some factors which you can control in your habitat — the use of poisonous sprays, for example — there are others, which may be less obvious.

Into this category fall those delightful furry creatures — cats. Not only are they responsible for the deaths of countless numbers of birds annually, but they will also attack other animals. In the case of actual bird numbers the activity of cats might not be so disastrous, because the birds are able to make up their numbers. The trouble is that cats are responsible for attacking and destroying nests and young as well. Quite obviously it is difficult preventing cats from working your patch, but there are ways of helping to protect the bird visitors. Care should be taken when siting bird tables, next boxes, *etc*, and it is also a good idea to have main feeding times. If you do this, the birds will soon get used to the idea and you can also keep roaming pets in during this time, and deal with any trespassers.

Wild Birds And The Law

It must be remembered that there are various Protection of Birds Acts, and even though the birds are nesting in your garden you are not exempt from this legislation. It must not be forgotten that birds, as well as their eggs, are protected by the law. The pertinent points from such legislation are listed below.

You must not:
● Take, damage or destroy the nest of any wild bird whist in use
● Kill or attempt to kill or injure a wild bird
● Take or destroy the eggs of any wild bird
● Ring or mark any wild bird (except in the case of an authorised person, for scientific purposes)
● Disturb any wild bird whilst on or near a nest or with young
● Have in your possession or control any wild bird recently killed.

A complete list of the legislation can be obtained from the RSPB (address on page 45). Please enclose return postage and something to cover the cost.

Planning The Bird Garden

If you think about the sort of habitats birds have in the natural state, you will soon realize that these hardly resemble the gardens to which we are accustomed. So perhaps the bird garden which we can provide, if it is really to attract and support birds, will not be a conventional garden at all. Since it is not the normal cultivated plants which supply the birds with their food, but uncultivated or wild ones, our garden will have to contain some of these too, even if it makes us just that little bit unpopular with our neighbours.

In the wild the birds will certainly range over a larger area than we expect them to do in our garden, where the plants which we are growing and the layout will be more concentrated. We will suggest some of the plants which can be grown in the table on page 00, but if you have room, you should try to encourage an almost impenetrable tangle of undergrowth in part of your garden: an area which the birds will enjoy, and which will not only allow them some degree of privacy, but which will, to some extent, keep people out.

Even if you have to live deep in the heart of inner suburbia, you've probably had a bird or two visiting your garden out of curiosity. Maybe it's already a place for birds, but the chances are that, if they are wise, the bird visitors will give your little bit of the urban habitat a miss next time they are around. Yet if you have something which will positively encourage them to visit, they are much more likely to return. If yours is a sterile jungle, the birds have been no more than curious, or perhaps have paused to rest, but unless there is something to attract them back they certainly won't make a regular habit of visiting your garden.

But wait, because you could turn even the smallest garden into a place which birds will visit, and it can be done quite easily, and with little expense. What you make of your garden as a bird reserve will depend on you. You could provide food and shelter in the form of bushes and trees and nesting sites. There are two ways in which you could encourage birds to feed. You could provide natural foods for them to eat by growing the right plants and shrubs, and you can provide food of your own making for the bird table. By increasing the range of food and by providing your birds with nesting boxes (see page 31) you may well find that the number and variety of species which become regular visitors will increase dramatically.

You will probably have noticed that whilst there are certain species which visit your garden all the year round − the blackbirds, sparrows, and the greedy starling − other birds only appear during the winter. This is because birds like the robin find food in the wider countryside less easy to obtain in winter, and move closer to human habitation for nourishment during adverse conditions.

Food Plants For Birds

Let's deal with plants which will provide birds with food in the winter. Although we give quite a large selection of species here, those which you can plant will depend on various factors, including the size of your garden, aspect, soil, *etc*. In the section on concrete yard gardening, we suggest species for growing in tubs when soil and space are at a premium. If you plant shrubs which will produce berries, not only will you have a possible source of food should the birds desperately need it, but for a short while at least you will have an array of bright berries to enhance the garden and cheer up the winter scene.

What sort of species should you plant? Birds will take most sorts of berries. In a good year hawthorns will abound with clusters of ripe haws, a true delight to many bird species (we deal with hawthorn in our section on hedges on page 00), and other species to include are cotoneaster, holly, viburnum and berberis.

Berberis (Barberry)
Has large numbers of both flowers and berries, especially when planted in full sun. This plant is equally at home, however, in partial shade, and can thrive in almost any soil as long as it is well drained. The plant can be propagated from hardwood cuttings, which should be taken in September. Dead and old wood should be thinned out in spring, but the species does not need to be pruned. Two evergreen forms are *Berberis stenophylla* and *Berberis darwinii:*
● *Berberis stenophylla* This grows to a maximum height of about 3 metres (10 feet) with a 4 metre spread (13 feet). The golden flowers appear in April. The berries appear later in the year. They are blue with a white bloom to the surface.
● *Berberis darwinii* There is not a great deal of variation in maximum spread and height, both of which are around 2.5 to 3 metres (8 and 10 feet). Flowers on this species appear in May and June, and are orange-coloured. The berries of a purple to blue hue appear later in the year.

Cotoneaster
There are a number of different species of this family of attractive shrubs. You will need to select

the varieties to suit your situation. There are evergreen as well as deciduous species. They are all extremely valuable in the garden, because of the large crop of bright red or orange berries which they produce. Many have particularly attractive foliage especially in the autumn. Since there is such a large variety of species, it is to be expected that there is also a large variety of form to match. Some cotoneasters are extremely low-growing: others may reach a height of more than 7 metres (23 feet). They appear to have no maximum growth of spread. All species have small

FOOD PLANTS FOR BIRDS

This table sets out some common shrubs which can be planted in the garden to attract birds, as well as some annual and wild plants. The list is not intended to be exhaustive and can be added to from your own experience.

Elder (*Sambucus nigra*)

Yew (*Taxus baccata*)

Cotoneaster simonsii

Contoneaster watereri

Cotoneaster horizontalis

Firethorn (*Pyracantha coccinea*)

Guelder rose (*Viburnum opulus*)

Barberry (*Berberis darwinii*)

Barberry (*B. rubrostilla*)

Barberry (*B. aggregata*)

Japanese laurel (*Aucuba*)

Wayfaring tree (*Viburnum lantana*)

Spindle (*Euonymus curopaeus*)

Autumn olive (*Elaeagnus umbellata*)

Russian olive (*E. augustifolia*)

Snowberry (*Symphoricarpos* spp)

Hawthorn (*Crataegus* spp)

Blackberry (*Rubus* spp)

Rowan (*Sorbus aucuparia*)

Holly (*Ilex aquifolium*)

Holly (*I. Mme Briot*)

Holly (*I. pyramidalis*)

Pernettyas (*P. mascula and P. mucronata*)

Cherries (*Prunus* spp)

Roses – Climbing (*Rosa* spp)

Honeysuckle (*Lonicera* spp)

Sea buckthorn (*Hippophae rhamnoides*)

Sunflower (*Helianthus* spp)

Scabious (*Scabiosa*)

Michaelmas daisy (*Asternovi – belgii*)

Cosmos

Antirrhinum

Wild plants include:
Thistles, grasses, knapweed, ragwort, nettles, poppies, teasels

rose tinted or white flowers, the blooms appearing in May or June. To keep the plant in good health the straggling branches should be removed by pruning during the winter. Several species are suitable for growth in the garden:

• *Cotoneaster horizontalis* is perhaps one of the most common. If you have a wall or trellis work to cover, this species will grow up to 3 metres (10 feet) against a wall. It has extremely rapid growth. The vivid red berries will appear on the shrub in autumn, and generally last well into the winter months.

• *Cotoneaster adpressa* is one species which is suitable for the rock garden. It is deciduous, and has a maximum height of 30 centimetres (12 inches). Red berries appear in the autumn.

• *Cotoneaster bullata* is much larger than the last species and has a maximum height and spread of about 5 metres (16 feet). The cherry-red berries appear on this deciduous species in autumn.

• *Cotoneaster frigida* has a maximum spread of about 8 metres (26 feet): height up to 7 metres (23 feet). It is useful for screening, and has the added advantage that it grows quite rapidly.

• *Cotoneaster microphylla*. It seems that this specimen spreads indefinitely. The maximum height is about a metre (3 feet). This species is evergreen and produces a good crop of berries in autumn.

One species which should be avoided is *Cotoneaster conspicuax decora*, mainly because its berries are not touched by birds. With the exception of this species the berries produced by the others will attract many birds, including finches, tits, thrushes and blackbirds. In addition the densely interwoven stems of many varieties provide fine nesting sites for such species as the spotted flycatcher.

Holly (Ilex spp.)

This popular, well-known evergreen plant is suitable for any soil, and will do just as well in exposed gardens in the country as in secluded town plots. Its only drawback is that it is slow-growing, but in spite of this it can reach a height of up to 16 metres (52 feet), having a spread of up to 11 metres (36 feet). The species can be propagated in August, by the means of half-ripe cuttings.

Holly does not transplant very easily, so care should be taken to include a ball of soil around the roots. This problem of moving plants can be overcome by purchasing container-grown specimens, and although the initial cost is slightly more, at least they are more likely to survive. The best time for moving them is in April, September or October. Holly does not need to be pruned. The most common species, and the one which grows in the wild, is *Ilex aquilifolium*. However, it is one of those species which is rather erratic as far as berry-producing activities are concerned, which means that there will certainly be years when there will be no berries at all. In cross-pollinating, and by producing cultivated varieties, as opposed to wild ones, one has a higher chance of a successful crop most years. It must also be remembered that each tree is of a particular sex – either male or female. Berries will only be produced when both trees are close by, so that cross-pollination can take place.

If you have a small garden you will probably favour one of the smaller variegated varieties, such as *Ilex golden king*. This grows to a height of just over 3 metres (10 feet), and has the advantages of producing extremely fine crops of fruit. Some species, like *Ilex Mme Briot*, have golden berries, and are suitable for the smaller garden, since they reach a maximum height of between 5 and 6 metres (16 and 19 feet). If you have a large garden, then you might want to try *Ilex pyramidalis* because, in favourable conditions, this species will reach a height in excess of 13 metres (42 feet). Although it has regular good crops of berries, its size might be prohibitive in some gardens. Some people are tempted to plant it, and then prune it drastically each year. This leads to a decline in the amount of fruit which is produced: in some cases it might stop berry production altogether.

Holly is also a very good hedging plant, as you will discover on page 135. The fact that it provides dense cover, with a good variety of nesting sites, as well as winter food, might outweigh the disadvantages of its slow growth.

Viburnum

This shrub is at its best where the soil is both rich and moist. Sunny positions are preferred by most species, although some do well in partial, but not complete, shade. The plant is reasonably easy to propagate, using heel cuttings, from half ripe wood. Such cuttings can be taken between June and August. There are both deciduous and evergreen species: the evergreen varieties need to be pruned regularly in April. The pruning of the deciduous species is much less frequent: old wood needs taking out from time to time. One species which is particularly suitable for the small garden is *Viburnum compactum*, better known as the cultivated form of the guelder rose. It reaches a maximum height of around 2 metres (6.5 feet), and the red berries borne on flat heads in the autumn provide a wealth of food. There is the native species *Viburnum opulus*, which is suitable for larger gardens, eventually growing upwards to 6 metres (19 feet). One other species which has yellow berries is *Viburnum anthocarpum*: its height is around 5 metres (16 feet).

Spindle
Unfortunately spindle (*Euonymus europaeus*), one of our native trees, is not as widespread as some of our other common species. Although referred to as a tree, it is generally seen as a shrubby bush, and does not usually reach a height of more than about 3 metres (10 feet). It is extremely poisonous and so should not be planted where young children are present. The fruits which will eventually appear vary in colour from a pink through to a more distinct orange.

Yew
This species (*Taxus baccata*) is generally used for hedging (see page 135), although there is no reason why it should not be planted as single species. It will provide an abundance of berries in the autumn, and so attract birds to the garden. The fleshy fruits are particularly relished by many birds, although the seeds are poisonous. One of our longest-lived species, some yews are said to be 2,000 years old. Reaching a maximum height of 10 metres (33 feet) they provide useful nesting places as well as food for some species.

Mountain Ash (Rowan)
If you have a reasonably sized garden you can encourage the mountain ash (*Sorbus spp*). It never reaches a great height – a maximum of around 8 metres (26 feet) is usually quoted, although some never grow to more than 5 metres (16 feet) – but it does have a large spread. During early summer your garden will be enhanced by the fine clusters of white flowers. Autumn sees the development of luscious red fruits, much sought out by seed-eating birds. The tree has been cultivated, and one of the species which is suitable for smaller gardens is *Sorbus vilmorinii*, with a maximum height of about 4 metres (13 feet). Pink berries appear in the autumn.

Honeysuckle
Still found wild in many tracts of woodland. Both wild and cultivated species of honeysuckle (*Lonicera*) will produce berries which are favourites with many garden birds, including the warblers and blackcaps. When well-established its intertwined stems provide useful nesting cover, with sites taken up by flycatchers.

Birds And Their Feeding Habits

Quite often when people want to encourage birds to feed in their garden they provide elaborate bird tables and, although the number and variety of birds might increase slightly, they are rather perturbed that many of the birds which they have read about as 'visiting gardens', just do not appear. Before you get disappointed it is well worth discussing some of the reasons why birds come, and why other birds don't visit the garden. As we said earlier the part of the country in which you live will influence the visitors to your garden. Although many species are widely distributed there are others which are found in some areas and not in others. Any good bird book (see page 45) will give you the necessary basic information. The type of food you offer, and the way in which it is given will influence the species coming for their fill. It must also be remembered that except in harsh winters many birds will take their nourishment away from your 'unnatural' garden, though there are always the opportunists, like the sparrows, tits, starlings, thrushes and blackbirds. Although some birds will feed almost anywhere many others are shy and these will need some cover for their feeding activities.

Perhaps one of the most important points to remember is the way in which the birds feed. Many species will come regularly and happily to your bird table, but there are others which prefer to scavenge their food from the ground. It must also be borne in mind that some species, like starlings, will deprive the smaller birds of their food, and you may find that all that you are doing is supplying an endless quantity of food for these quarrelsome, greedy birds. Arrangements must be made for dealing with this. It is important to take into consideration the different ways in which birds feed, when planning your feeding assignments. There are those which feed on the ground, those which prefer a high perch, and some

Starling

which are so shy that they need the shelter of a hedge if they are to come to the garden at all.

Although we have indicated that the starling is likely to devour food put out for other birds, there are ways of overcoming this problem. As well as being greedy, often aggressive visitors, starlings are rather lazy birds who often leave their roosts after other birds have already been to the garden and taken their fill. Earlier too than many other species, the large, noisy flocks will make tracks for their sleeping quarters. To some extent the problem of feeding is therefore solved. If food is offered early in the morning and/or late at night you will make sure that other species get their share. Some observers tell us that the starling, along with the house sparrow is not fond of human companionship, and both species prefer to feed away from the house. I'm not so sure that I altogether agree with this, but nevertheless if you have a large enough garden and can have a feeding station away from the house you might deter them from upsetting the smaller birds.

Just as we humans benefit from a regular meal, birds also like a feeding timetable. Whether you scatter the food or place it on the table it is a good idea to ensure that, as far as possible, they are fed at the same time each day. To give the greatest benefit your aim should be to make sure that food is placed in your garden early in the morning. Birds tend to lose some weight over-night, and by having a good start they can replace this, ready for an active day ahead.

Bird Tables And Feeders

If you are a handyman you will want to make your own bird table. If you are not you will need to purchase one (see page 46). It has already been mentioned that birds feed in different ways, and this fact must be borne in mind when making arrangements for buying and siting feeders. Of course the simplest thing to do is to throw the food onto the ground. Within seconds of doing this, wherever you live, you will soon have one or two species of birds investigating the material. It depends on the food being offered as to whether all of it will be taken. If some is left it might encourage mammals, like the rat, to come and take what is left after darkness has fallen. If you do not want to go to the lengths of either making or buying a bird table, you can place your food on a tray which you bring in at the end of the day.

Many people want to encourage birds closer to their houses. Unfortunately many modern houses have window sills which are not suitable since they are too narrow or slope downwards at too great an angle. Broader window sills are suitable places, since food can easily be put out when conditions are right. There is no reason why you should not attach a bird tray to your window ledge, so that you can get a good view of your feeding birds. One firm supplies window-ledge feeders (see page 46). With the birds close at hand you will soon find that there are characters which you get to know.

It only stands to reason that a properly erected

17

bird table will be more useful and will save you the trouble of having to carry in a tray every night. The simplest way to make a bird table is to secure a wooden tray to a post. A piece of seasoned wood about 15 x 15 cm (12 x 12 in) is attached to a 5 x 5 cm (2 x 2 in) wooden post, using metal brackets. A narrow rim, about 1.5 cm ($\frac{1}{2}$ in), is attached all the way round the edge of the tray. This stops food either being pushed off by the birds, or being blown off by strong winds. The stake can be sunk into the ground. Alternatively you can dispense with the stake and attach the tray to either a 'T' or an 'L' bracket which in turn can be attached to a wall. Another alternative is to attach chains to the corners of the table so that the whole can be suspended from a branch of a tree. This is a basic bird table, and it does have its drawbacks. There is no roof, which means that food will soon get wet, and this is particularly a problem in snowy weather. You can make a cover by attaching four posts to the corners to support a sloping roof.

Protecting the bird table

There are other problems with the tray-type feeding platform, since it is vulnerable to cats and squirrels. Many ideas have been tried out to prevent these

Nuthatch

animals from taking advantage of the food at the table. Of course one could put barbed wire up the post, but this is likely to be dangerous to both animals and young children. A piece of guttering pushed up over the stake goes some way to stopping cats and squirrels, as they are usually unable to gain

Robin

a grip on the slippery surface. If your table is not very well positioned then cats can by-pass the drainpipe by springing at the table from a distance, and will usually make it if the table is not too high. Once squirrels have discovered a new free source of food they will go to great lengths to make sure that they can get at it. I have seen a tin, like that used for biscuits, attached underneath the table to ward them off. Various firms supply and sell bird tables, and these vary from the tray type to rustic pieces, aimed at fitting in with the most olde worlde garden. Whilst the latter style is very attractive, the post on which the table is supported is ideal for cats and squirrels to climb. Details of the suppliers of tables are given on page 46.

Siting the bird table

It should be remembered that the siting of the table is of great importance. Firstly, it is no good suspending it from a bough if cats or squirrels are perpetually in the tree and frighten the birds away. Then, while there are some bird species which are almost extrovert by nature, like the house sparrow and the starling, there are others which are more cautious, preferring not to exhibit themselves to the same extent: those frequent visitors both to the garden and the bird table, the blue tits, come to mind. When siting the bird table this should be borne in mind. If you position your table in the middle of a wide open space the chances are that certain species will always avoid it, or at least not use it as much as if it were easy to get to. If there are bushes and hedges in your garden which can act as stopping-off points, leaving a short distance to the table, the blue tits will come more easily and use it.

Nightingale

Fair shares for all

One of the other problems which you are bound to encounter is the speed at which food disappears before very many birds have managed to get a look in. Some people have suggested placing a layer of fine mesh wire netting just above the food, so that the birds can peck it through the holes. If this is attached to the table it poses problems for cleaning, and to some extent, when putting out food. Far better to make a frame which will fit inside the bird table, and then cover this with wire netting, so that it can be removed. To make it possible to remove it easily, two staples can be hammered into the frame, so that you have small handles for lifting it out.

It is worth bearing in mind the size of the material being offered to the birds. If the food is very large the birds which might want to make off with it will not be able to do this. Very small pieces of food are preferable as they discourage birds from going off with beak-fulls. The Royal Society for the Protection of Birds produces a very versatile bird table, to which a seed hopper can be attached. This is useful because, placed under the roof, the seed is kept dry, and if added to the material provided for other species, seeds will encourage the seed-eaters. You can also screw in cup hooks and suspend various small baskets and feeders from your table, although you might want to distribute the material over a wider area of your garden, especially if you have room.

Small bird feeders

Over the years various firms have produced a variety of small bird feeders. The simple nut bag, sold by many shops, is useful, except that the constant

Thrush

19

SOME COMMON BIRDS AND THEIR HABITS

This table shows a variety of birds which are garden visitors in many areas of Britain and also sets out their average longevity, when they are likely to be seen, plus their feeding and nesting habits.

KEY: BIRD TABLE Y = Yes N = No
      NEST BOXES S = Sometimes F = Frequently N = Never
      SEEN R = Resident SR = Summer resident WV = Winter visitor

SPECIES	LIVES TO	SEEN	FOOD	USES BIRD TABLE	USES NEST BOX
BARN OWL	15	R	Small rodents and birds	N	S
BLACKBIRD	10	R	Wide range – invertebrates (worms/insects), seeds berries, fruits	Y	S
BLACKCAP	5	SR	Insects, fruits, berries	Y	N
BLUE TIT	11	R	Seeds and insects	Y	F
BULLFINCH	8	R	Mainly weed seeds, with some berries and buds	Y	N
CHAFFINCH	10	R	Invertebrates – insects/spiders, buds/fruit	Y	N
COAL TIT	6	R	Invertebrates (spiders/insects) – preferably from pine trees. Also some nuts and seeds	Y	F
COLLARED DOVE	5	R	Corn, some berries, buds	N	N
DUNNOCK	8	R	Usually insects during summer: weed seeds in winter	Y	N
FIELDFARE	5	WV	Berries and invertebrates – insects, slugs, spiders	Y	N
GOLDCREST	3	R	Invertebrates – insects and spiders	Y	N
GOLDFINCH	8	R	Mainly seeds, including thistles: will take insects	Y	N
GREEN WOODPECKER	5	R	Insect larvae from trees	Y	S
GREAT TIT	10	R	Insects, spiders, seeds and nuts	Y	F
HOUSE MARTIN	6	SR	Insects	N	S
HOUSE SPARROW	11	R	Wide range – grain, insects, other seeds	Y	F
JACKDAW	14	R	Variety of plant and animal material	Y	F
LITTLE OWL	16	R	Insects, small mammals; sometimes birds	N	S
LONG TAILED TIT	7	R	Insects and seeds, almost always taken in a tree	Y	F
MAGPIE	15	R	Insects and small animals, including birds	Y	N
MARSH TIT	10	R	Seeds of weed plants, berries and insects	Y	F
MISTLE THRUSH	10	R	Wide range of wild fruits, insects and larvae, earthworms and snails	Y	N
NUTHATCH	9	R	Mainly nuts, including mast and acorns, some insects	Y	F
PIED FLYCATCHER	9	SR	Insects caught whilst in flight: also taken from trees and tall bushes/shrubs	N	F
PIED WAGTAIL	7	R	Almost exclusively insects	Y	S
REDSTART	7	SR	Mainly insects	N	F
REDWING	19	WV	Invertebrates, including insects and worms	Y	N
ROBIN	11	R	Wide range of insects, spiders, seeds, berries	Y	F
SONG THRUSH	14	R	Invertebrates, berries and seeds	Y	N
SPOTTED FLYCATCHER	8	SR	Flying insects	N	F
STARLING	20	R	Almost anything – seeds, insects, plant material, etc	Y	F
SWALLOW	19	SR	Insects caught on the wing	N	S
SWIFT	21	SR	Insects taken while flying	N	S
TREE CREEPER	7	R	Mainly insects	N	S
WREN	5	R	Invertebrates – insects and spiders	Y	S
YELLOWHAMMER	7	R	Seeds, fruits and insects	Y	N

Finches

Wren

pecking by some birds does damage the net, and the nuts fall out. One cannot do better than buy equipment sold by the RSPB, or advertised in *Birds* Magazine. The names and addresses of producers/manufacturers/distributors are given on page 46.

From experience you may need to move your feeders around. If you find that few birds come to feed then it could be in the wrong place. Don't forget, however, that you will need to leave it in each position for a while until the birds actually discover it. You might decide that a couple of weeks is long enough, in one spot. It is a good idea to move your feeder or feeders at least once during the feeding season.

If you have access to some small logs, you will find that these can be made into useful feeding 'baskets'. Drill holes about 4 cm (1½ in) in diameter right through the log. You can then make either a fat mixture, or a cake, as we will suggest, and fill the holes with this. Once hung up this will be eagerly picked at by a number of species. If you knock a staple into the top of the log you can tie string with which to suspend it around this. These days many different varieties of convenience foods arrive in the house in an assortment of containers. Most of these are made of plastic. They too make useful food containers. The advantage of these containers is that you can keep a supply handy, and fill them with food material. Most will withstand quite hot materials. Using a large darning needle and string thread

hangers through the container, so that they can be hung from hooks or branches.

Making your own feeding bags
You will see peanuts sold for birds in those small string bags, but you can make your own for nothing and cut out this cost. In many supermarkets nuts, as well as various types of fruit, are now packaged in small net-like plastic bags. Once you have taken out your fruit, make sure that one end is securely tied. Now fasten a piece of string to the other end, but before tying, fill it with nuts. Suspend the bag from a branch, hook or bracket and birds will take their fill readily.

Supplementary Bird Food

To encourage a wide variety of birds you will need to provide differing types of food, apart from the seeds and berries your garden produces. Late autumn to late spring is the period when birds are most likely to have difficulties in locating supplies and may starve. During the rest of the year whether food should be provided is arguable. There are some ornithologists who definitely say *NO:* there are others who are of the opinion that since birds come to look for a supply of food in your garden you should provide them with a source all the year round. There are dangers in this view, because there are some foods which birds might eat in the winter, and continue to take in the spring, which they may try to feed to their young. This is particularly true of peanuts, for example. The adults relish them and eagerly peck at the nut bags hanging in the garden. But the young cannot digest them. It is to be expected that if you provide a wide variety of food for the birds in winter the numbers of species visiting your garden will increase. If you stop this in spring, the number and variety might decline. They would thin out to some extent anyway, leaving to seek out other areas where they can obtain their food naturally. We suggest that it is best to decrease the amount of food offered during the second half of March, and that it should finally cease by the end of the month.

It must be remembered that birds can be grouped according to the way in which they feed. There are those which feed on seeds: others take worms and snails, and there are more which are omnivorous – feeding on a mixture of seeds and flesh. For

GOD IS LOVE.

PEACE ON EARTH.

GLORY TO GOD.

GOOD WILL TO MEN.

ROBIN'S PLEA.

Little streams become flowing rivers. "ROBIN DINNERS" are becoming quite a national institution. In London alone nearly Twenty Thousand "human Robins" are now every year "made happy for an evening." Robin wants to have a "*Robin Dinner*" in every town and village in the land. Send a penny stamp addressed to "*Robin*," 7, *The Paragon, Blackheath, S.E.*, and he will tell you how it is to be done.

Kind words, and loving deeds, and tender sympathy, are gifts all can bestow: and these at Christmas-tide should be scattered everywhere.

Robin wishes you, one and all,—

"A Holy Christmas, and a Happy New Year!"

23

example, you will almost certainly have seen the thrushes in your garden as they have paraded the lawn looking for food. If you have a hedge you will probably have found banded snails taking shelter there: here the thrush will come for a good supply of readily available food. If a large stone is placed by the wall, it is possible that the thrush will use it for its anvil – a stone on which the thrush bashes the snail shell to get at the soft-bodied animal inside.

Table scraps

Many of your table scraps will provide very useful material for feeding to the birds. One of the items which is normally thrown out to birds is the stale bread. If you throw this on the lawn, birds like sparrows and starlings will quickly swoop down and take it. Seldom do you see other species, except perhaps the occasional blackbird. *We ought to stress that bread, and particularly the white variety, does not have great nutritional value.* By continuing to give bread only you will be catering for only a few species, and they will have to search out other food to make up the balanced diet which they need. By offering nothing but bread you are thus eliminating more interesting birds.

If you start to collect together a variety of kitchen scraps and feed these to your birds you will soon see the difference. Hang the carcase of a chicken from the bird table or a tree, and watch the antics of various species as they try to clean the bones. If you have an old mug, fill this with fat left over from your cooking – especially from meats. Hang this with string from the branches of a tree or near the bird table, and you can watch the tits balancing on the edge, picking out the fat to their hearts' content. But be warned, for the starling is a very crafty bird, and within a short time, it too may soon learn to perch on the cup and get at the fat.

Suet provides a great deal of nutrition for birds. If you push small pieces into cracks in trees or around the bird table, you will find that several species will soon seek it out. Stale cheese and the cooked rind from bacon, as well as bacon itself, is much relished. There are other 'wastes' from the kitchen: instead of throwing cake away when it gets stale, you can give it to the birds. Many households usually have a few potatoes left over and these can be placed on the bird table. If you are having baked potatoes place an extra one in the oven, and your birds will soon make light work of it.

Crows in winter

Bird 'puddings'

You might like to make your own puddings, using fat, such as lard, as this is useful for binding various materials together as well as being nutritious. Many natural ingredients can be bought cheaply from wholefood shops – oatmeal, various seeds, nuts and peanuts, for example – and these are much more valuable to birds than those which have undergone processing. To these one can add table scraps, making a very good cake for the birds. Use about twice the volume of scraps to fat. Scraps from the kitchen can also be put into small tins and if hot fat is poured over them, it will bind all the materials together. When the mixture has set it can be turned out. Empty coconut shells can be filled with scraps or fat and suspended from a suitable hook. You will find it easier to put a string through the holes before putting in food. It is also possible to buy tit bells, and you can fill these with food as well. If you want to prepare a regular cake and not rely solely on kitchen scraps, you can make one of these when the weekly baking is being done. It is worth purchasing wholemeal flour, as this has better nutritional value. About 1kg (2lb) of flour should be mixed with 225g

24

Sparrows in the snow

($\frac{1}{2}$ lb) of margarine by rubbing in with your fingers. Then stir in water until the mixture is quite thick: this should then be baked in a cake tin.

Commercial bird food
You can of course buy bird food, and the advantage of buying a proprietary brand is that the packet contains a balanced diet for the birds, with a wide variety of seeds ensuring that seed-eaters with differing tastes are catered for. While this is good for the birds, naturally the 'scientific' formula and packaging mean that the mixture works out quite expensively. Furthermore, if you are in an area where there is not a great diversity of species you will often find that quite a lot of the seed mixture will remain uneaten. It will always be the case that while birds may visit a garden in one area, they will stay away in other places, because they have a regular supply of natural food, and don't need to take advantage of your hospitality.

If you think packaged brands are rather an expensive way of buying bird food, you will be able to purchase seed mixtures from most pet stores or in large quantities by post. Amongst the seeds which

you could buy to mix for yourself are the old favourites like millet and sunflower. Add oats to these and you will soon have the birds falling over themselves to feed! If you are lucky enough to have visits from nuthatches you will find that they greatly relish hemp. If it is left as it is the nuthatches will take it, but if you buy it in crushed form you will find that many smaller species will enjoy it too.

Peanuts and coconuts
We have omitted what is probably the most common bird food and undoubtedly one which will be taken regularly, namely peanuts, which are very valuable as a bird food, being particularly nutritious. There are various ways of presenting these to your birds. Tits will be happy if you place shelled nuts in a tit feeder. It is worth spending a while making strings of peanuts, because the antics performed, especially by members of the tit family, and in some localities by nuthatches, are certainly worth watching. Coconuts and nuts are a favourite delicacy, but the price of these has increased considerably. If you do buy a coconut, don't leave it as it is, but saw it in halves, suspending the pieces from a bird table, or some other suitable site. *Never* give birds desiccated coconut, because as this is dried, as soon as the birds have swallowed it the liquid in the stomach will make it increase in size, and it could prove lethal.

Fruit
You will probably find that your local fruit shop or even supermarket will sell its ripe fruit at greatly reduced prices. Some people even have arrangements with their greengrocer for them to save overripe fruit, which they will often sell at a reasonable fee. Oranges, apples and bananas are all of value. The skins need to be removed and the fruit cut into suitably sized pieces. You will find that while some birds are content to eat the food at the bird table, there are others which prefer to eat in private, and will carry off pieces of the fruit.

Ants' eggs and mealworms
So far we have only dealt with birds which feed on seed and soft materials. Even in very cold weather these birds can usually find something to eat in the wild, although those species which rely on, for example, worms, will suffer when the ground is frozen. To satisfy the insect eaters there are two good buys. There are ants' eggs which are normally

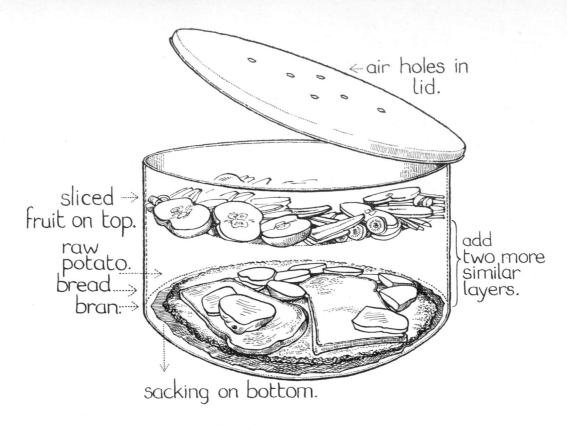

air holes in lid.

sliced fruit on top.

raw potato.
bread.
bran.

add two more similar layers.

sacking on bottom.

Cultivating a meal-worm colony, as described on this page

fed to fish. Of course you could do your own collecting, and both eggs and larvae – which you are likely to encounter under stones – can be fed to your birds. Don't forget that there is other life under the stone, and stones should always be replaced carefully to prevent any animals living there from being damaged. The other easily obtained insect food is the mealworm. You could keep your own culture once you have bought some. If this does not appeal to you, regular supplies can often be purchased from petshops and fishing tackle suppliers.

To prepare and maintain a mealworm colony you can use either one of the old globe-type goldfish bowls or a large circular tin. Although the goldfish bowl does not need a lid, the tin does. Using a hammer and nail, punch some holes in the tin lid, to provide ventilation. A piece of sacking – narrow mesh – should be placed in the bottom of the container. Using a medium-coarse bran, place a 5 mm (0.2 in) layer on top of the sacking. You will now need a couple of slices of bread and some raw potato, which are placed on the bran. Repeat this twice more, so that you will have three layers, then on the top of this place some fruit and vegetables,

taking your choice from sliced apples, carrots, *etc*, depending on those which are in season. Once this is set up you are ready to introduce your mealworms. Purchase them from a local supplier: your first batch should contain between a hundred and two hundred individuals. They will go through the normal insect life stages. When you buy them they are in the larvae stage; in three weeks or so they will change into pupae, and these will eventually give rise to the beetles. The cycle will now be repeated as the beetles lay their eggs. Successful breeding depends on a number of factors, including correct ventilation and warmth. The container should be kept at normal room temperature. If for any reason the colony gets damp the individuals are likely to die.

Collecting from the countryside

For the person who wishes to cultivate the garden as a wild bird sanctuary, however small this might be, and provide food in the worst months of the year, autumn is a very important time. Autumn is the period of seedtime and harvest, so plan for your own needs and those of your feathered garden visitors. As with your own fruits those which you collect for the birds need to be carefully stored. Only sound berries should be selected, and any which show signs of deterioration should be fed to your birds immediately. Before storing your supplies in shallow boxes in a dark place you must make sure that berries and nuts are carefully dried. If you do not do this, by the time that the birds need them, which is often much later in the winter – around February – they will have rotted.

You will need to keep a careful watch on the hedgerows because, like fruits for your own use, the berries and nuts for the birds need to be collected at their best. This also helps to ensure that they keep well. Depending on your location the first fruits will probably be ready as early as August. The mountain ash (rowan) produces a very good berry, much relished by birds. Most years will see crops of elderberry, and if you have any of the fruits near you you will quickly realize how much birds take them from the decorative purple stains on your car and paths. In some areas there are good crops of the fruit of the hawthorn – the haws, particularly in farming areas where the hawthorn was planted at the time of the enclosures to act as a field boundary. The blackthorn will usually produce good crops of sloes, the ancestor of our modern plum, but unfortunately birds will only take these when they are well ripe on the shrubs, so they are probably best left for the birds to help themselves. The country dweller still seeks out crab apples to make preserves: don't forget to get some for the birds, as they will peck at these to their hearts' content. Once the supply of food starts to dry up in the wild, you may even find fieldfares making sorties into your garden, especially if they find crab apples are about.

Later in the autumn many varieties of nuts will be available for the picking. The fruits of the horse-chestnut tree – the familiar conker – are much sought after, and these can be collected along with acorns, beech mast, sweet chestnuts and hazel. Those with softer outer shells need collecting as soon as they fall from the trees.

You can collect seeds too, and in a good season can find a plentiful supply particularly from knapweeds, ragwort and teasels, and you will probably have seen birds eagerly devouring these. *Don't* take them all from one area, as you will deprive the birds which don't normally come to the bird table of a valuable supply of food. The common stinging nettles should not be overlooked as they will also provide a useful supply of seeds. Where you have coniferous plantations you will be able to collect fallen cones. Some birds, like the crossbill, will usually take out the seeds themselves, but there are other species which cannot do this. Although it might be a tedious task to extract the seeds the value to the birds will far outweigh the tedium. These seeds need to be stored so that air can get at them so that they will not rot. Polythene bags aren't really suitable for this, and it is suggested that small muslin bags are made. Hang the seed bags in an outhouse or garage to store them and feed to the birds as necessary.

Providing Water

Water is, if anything, more important than food to the birds. Not only do they use it for drinking: they also revel in it when bathing. We suggest you make pools (see page 98), and pools with a shallow area are advisable so that birds will be able to use them for drinking purposes or for bathing. It is strange how the once-popular concrete bird bath, once a feature of so many gardens, seems to have vanished, and yet, with the piping of springs and drains, together with the filling-in of ponds, water for birds is becoming less and less available. The need for water varies from species to species. Whereas some birds, those which feed on moist, soft food – worms for example, will get a lot of their water from their food, the seed-eaters will get very little. Some species seem to be able to manage for a long time without water; others would die quite quickly if they didn't have a constant supply. It is almost certain that during harsh weather many birds die, not from starvation as would be expected, but from thirst.

Water in winter
Providing water in frosty weather is something of a problem, so perhaps it is as well to deal with this first. *Never* add glycerine to the water, because the birds will get it in their bills and on their feathers, and undoubtedly perish. You could build a small brick 'stove'. Two bricks high by two bricks wide will be suitable. A *shallow* container which will fit well should be placed on top. To keep the water free from ice all you will need to do is to place a night light underneath.

One of the easiest ways to keep water free from ice in a bird bath is to buy an aquarium thermostat. If you place this in the water container, and cover it with some gravel, it should successfully prevent the water from freezing. Cable used outside needs to be of the type meant for this particular job, and connectors should also be of the correct type. It is wise to get a qualified electrician either to check your installation, or to do the work for you – better to be safe than sorry.

Water for bathing

Being able to watch birds bathing in your garden will give added pleasure to your wildlife sanctuary, and you will undoubtedly learn a great deal about birds' behaviour from their bathing activities. For feathers to work efficiently they have to be kept in tip-top condition, and one of the ways in which birds maintain them is to take a bath. If you get a chance to watch birds in the rain you will see how they react. If they are out in a heavy shower, they position their bodies in such a way that the excess water runs off the feathers. Yet in a light shower they are likely to let the rain fall on their feathers, and then preen themselves, and I've just been watching a thrush do just that on a branch at the bottom of the garden!

Most bird baths tend to be on or near the ground, and naturally when the birds are bathing they become very preoccupied, and are likely to forget about possible predators. A bird bath high above the ground will solve this problem, although of course one cannot control a stealthy cat which could make an unwarranted attack on a bird either bathing or drinking. An accidental dip in the water by a marauding cat will probably deter its future escapades.

Make your own bird bath

If you don't have a water area, then you can make a simple one in your garden, with the emphasis especially on the needs of your birds.

Dig out a suitable shape, making sure that all sharp stones have been removed if you are going to use some form of thin liner. So that birds can get into the pool and drink, it is suggested that it slopes downwards from a minimum depth of about 25 mm (1 in) to about 30 cm (12 in). Such a gradual slope, over a length of about 1 metre (3 ft) will give you enough variety of depth to suit the various birds which, hopefully, will take advantage of your water bath.

Place some soft material – paper, peat or sawdust (check for sharp pieces of wood in the latter) into the hole to a depth of about 50 mm (2 in), so that it will form a cushion for your liner. Carefully place the liner in position and use stones or turf to hold the edges in place. It is a good idea to grow some aquatic plants in your pool. You can either plant these in pots or you can put soil directly onto the polythene. If you take the latter course of action you will need to place a small sheet of cork or other suitable material onto the liner, so that when you pour in the water you will not stir up the soil.

If you feel that you either can't go to the trouble of making a pond or you don't have the space, the dustbin lid technique will suffice. Beware of rubber lids, because water often causes the surface to come off, and this could damage the birds. For safety use either plastic or galvanised types and wedge them securely. The value of your dustbin lid bird bath will be enhanced if you can manage to secure a branch of some sort in the lid. If you push a branch through the hole in an earthenware plant pot, and then upturn this, you can fill the container with soil or cement. The end should be sealed with polythene. The potted branch will serve as a natural perch for the birds.

If shallow bathing containers are used, they will quickly become fouled up, so empty the contents daily, and use a weak solution of disinfectant about once a week. This will kill any fleas, but it is important to make sure that there are no remains of disinfectant, before the container is replaced.

Making a dust bath

In particularly dry weather you've probably noticed birds taking a 'dip', not in water, but in dust. For some species this dry bathing seems to be as important as a water bath is to other birds. Apart from providing a service for your garden birds, the dust bowl will also give you pleasure as you watch the birds using it.

A box about a metre (39 in) square and 5 cm (2½ in) or so deep is about the right size. If you don't have a box this size, you can use a large plastic bowl – the square variety rather than the round ones – cut to size depth-wise. The advantage of using a container is that it can be taken in when wet weather threatens. If you have a large garden, there is no reason why you shouldn't build a permanent bird dust bath. A few narrow pieces of wood sunk into the ground will suffice. The receptacle needs to be filled with dry soil, and should be in such a position that it catches the sun. There is one problem which you are likely to encounter, and that is of course the rain. However, if you use a good soil mixture, and have the bath in a sunny position, the water is likely to drain away fairly quickly and the wind and sun will soon dry the contents. You could drill some drainage holes in the bottom of the plastic bowl if you are using one as the basis of a bath.

Some dry soil is needed, and this should be sieved to remove all the lumps and other impurities. You could perhaps add some sharp sand, and if you are lucky enough to live in a house with a coal fire, you could substitute ash for part of the soil.

One of the reasons why birds come to a dust bath is to get rid of some of the many parasites which cling to their bodies. Although many parasites die once they have parted company with their host, another bird or perhaps an animal may pick them up before this happens. Although it is difficult to eradicate the parasites, if you sprinkle small amounts of pyrethrum powder in amongst the soil mixture from time to time you will at least keep down the numbers of unwanted guests, without doing any harm to the birds.

Cats are very inquisitive animals, and may be tempted to use the dust bath, just as they use sandpits in the garden, for their toilet. A cover, particularly at night, will help to solve this annoying problem, and also perhaps keep out any unwanted dogs as well.

Nesting Sites And Boxes

As we have already mentioned, most people have small gardens which may seem to have little to offer birds in general. We have already discussed ways of attracting birds by feeding them. It is also possible to provide artificial nesting sites in the form of nest boxes. This is being done not only in more and more gardens, but even in the sort of places one finds birds nesting naturally, such as areas of woodland. The reason for providing nesting boxes in the wild is that many of the specialized requirements which certain species have are fast disappearing and so alternatives are necessary if the species is to survive.

Assessing nesting requirements

Before we look at the use of artificial nesting boxes, we need to know something about the nesting habits and habitats of some of the more common birds likely to visit the garden. If you happen to live in coastal areas you might already have unwanted

Missel thrush's nest

29

Nest of a yellow hammer

visitors in the shape of gulls nesting on your rooftops. They have learned to take advantage of the man-made habitat, miles from the nearest coast. It is likely that these birds never have seen, and probably never will see, the sea.

When looking at the subject of nesting sites we must remember that different birds have very different requirements. If you have seen nest-boxes advertised (and we offer some addresses of manufacturers/distributors on page 46), you will usually notice that they have holes in the front providing for those birds which nest in holes in trees. Not so very long ago the natural woodland would have provided birds with plenty of nesting sites, for as trees get old, they become less resistant, and the birds could either use holes already there, or make their own. Today, unfortunately, we tend to tidy up the countryside, and furthermore, because of our need for more and more timber, trees are not normally allowed to reach such a state. This is one of the reasons why we find many artificial nest-boxes in natural woodland.

There are other birds which do not need the seclusion of a hole in which to nest. The nests of the latter vary from a mere scrape in the ground to rather more elaborate and carefully excavated struc-

tures. In the garden you will usually find that it is the various tit species which take to the holes. Nuthatches might also take advantage of your hospitality and both redstarts and tree sparrows are also hole-nesting species. Of course, the prime example of hole-nesting birds are the woodpeckers and if you live in a suitable area you might find that these will occasionally come to your garden. Of the other groups of birds, those which prefer to build their own nests in the open, the thrushes, blackbirds and robins, are likely to be the most common visitors your patch for this purpose. It must be borne in mind that whether the birds are the sort which nest in holes or whether they have other requirements they need a degree of protection, particularly from the rain and the wind. For obvious reasons, such as the possibility of an attack from predators, birds will seek out a hidden and secluded site when building their nests, although quite often it is very close to human habitations.

Perhaps it would be as well to look at the natural sites before we deal with artificial ones. Shrubs, both single species and those planted in hedges, and trees naturally provide suitable places for birds to build nests. The make-up of a particular hedge will determine, to some extent, the usefulness of the

30

structure as a nesting place. Yew hedges are dense and provide innumerable places. The same is true of species such as holly, hawthorn and beech. Mixed hedges are also valuable, although there are some shrubs, and hazel comes to mind, which are not as useful to birds as other species. If you have looked carefully at the position of a nest in a hedge, you will see that the bird needs a base, on which to build its nest. Most nests are built where the branches are forked and hazel does not provide a large number of forks. When pruning your trees, you can ensure that you cut the twigs in such a way that you will leave a good supply of forks. Pruning should take place out-of-season, that is, during autumn and early spring, so that nesting birds are not disturbed.

Creepers are valuable placed against certain walls, because most carry a useful supply of food later in the year. Creepers are also very much sought after by various species of birds, and provide useful hibernatory sites for other animals, including mice and butterflies. No-one is really certain why a bird finds a particular location suitable to its needs. But rest assured that once it has chosen its site it will be satisfied. It is always tempting to try and help the bird improve its home by trimming the shrub or creeper, so that it is easier for the bird to get at the nest, or to push over some of the vegetation because it is felt that the nest site is too exposed. Please refrain from helping the bird, because it knows best and the chances are that you will hinder it and even perhaps cause it to leave and seek another nesting place outside your garden.

Making a nest-box
Having provided the natural sites, we now need to look at ways in which we can provide nesting-boxes. The term 'nest-box' covers all the man-made types, which range from small, closed boxes to simple open-plan trays. The type which you make will depend on your garden. Many of the boxes produced serve the same purpose. A glance through one or two of the catalogues produced by firms supplying nest-boxes will show a variety of types. The basic difference is in the size of the entrance hole. We have given only one design for a nest-box, since this is the easiest of the hole-type to make, and is based on a

Nest of a song thrush

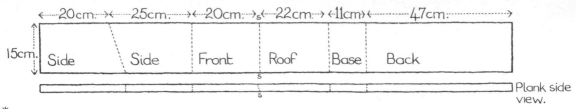

Plank side view.

* Make a slanting cut at "s".   For alternative open-fronted box – cut along line 'A'.

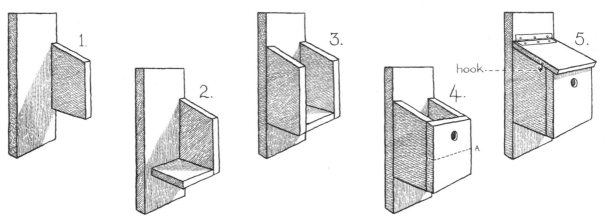

Making a nest-box, as described below

well-tried design, which appears in many books and leaflets about birds. You will need to decide what wood to use, and some care should be taken with this. Obviously hardwood is likely to stand up to the weather for a longer period of time, and is to be recommended for this reason, since it is reasonable to assume that once you have produced your nest-boxes you will want to use them over and over again for many years. As we shall see later, siting the nest-box is important, particularly in relation to changes in temperature; but fluctuations will still take place over a period of time, and a thicker wood will absorb these changes to a greater degree than will thinner material. If you don't want to buy new wood, you may find a demolition site near you where a contractor is burning small quantities of wood such as that taken up from flooring which he could be persuaded to give to you instead. The size of floorboards is likely to be about the same as that suggested in our diagram. A demolition merchant will have second-hand timber for sale very cheaply, and you may also be able to obtain other useful items from him. Old sinks, which can be made into bird baths, can often be found for nothing, since they are frequently thrown onto the rubbish dump.

If you decide to use softwood for your box, you

should go for one of the stronger timbers: cedar makes a good buy. Whether choosing soft or hard woods ensure that the wood has been seasoned properly, otherwise warping will occur, and the box will soon develop gaps. Before placing the nest-box in position, give it a coat of creosote to weather-proof it. Ensure that the creosote is dry before putting up the box. Some people suggest that it is a good idea to place a piece of roofing-felt on the top of the box. Care should be taken if you follow this suggestion, to make sure that the water doesn't trickle down this off the roof and into the hole.

In catering for nuthatches, tree sparrows and blue tits a hole about 28 mm ($1\frac{1}{8}$ in) in diameter is suitable. The position of the hole is of importance. Make sure that it is near the top. If it isn't cats may manage to attack and remove nestlings. And the feline members of the household may not be the only predators: other offenders may see the birds as they stretch their necks upwards in their hungry search for food, if the hole is too low. Of the whole opera-tion of making the nest-box it is making the hole which usually proves the most difficult. By far the easiest method is to use either an electric drill or a brace and bit. You can make a hole with the smaller drill, and then use a round file to enlarge it. There are other methods as well. By sawing the front down the middle, and then cutting out two semi-circles you will have your hole. The two halves of the box are joined together with either glue or wood braces nailed through from the front – make sure that the two halves fit together properly.

Once you have all your pieces you can assemble them with either screws or nails. If you are particularly adventurous and use dovetails the joints will need gluing. Screws have a distinct advantage over nails, since you might want to take the box to pieces at some stage. To ensure that the whole structure is waterproof, seal the joints with a compound such as Bostik or Seelastic. When fixing the top with the hinge you should make sure that it fits firmly against the body of the box, otherwise water will seep in. In the natural state many birds die because water gets into their nests. You don't want to be a party to this tragedy in your garden. A catch must be fixed on either side of the box so that the lid is held firmly in position. Some people prefer the 'lid' to be on the side, but rain is more likely to drive in if the box is made this way. One way of preventing this is to fix

wood struts inside the box, so that the door fits against these, rather like the door in a house. If you are a good workman and manage to make the box to a high standard, you will need to drill a ventilation hole or two, as well as drainage holes in the bottom of the box. Although some birds do remove their own and their nestlings' droppings from the nest, others do not, and a drainage hole, whilst not solving the problem completely, will help with hygiene!

If you live in an area where the birds already mentioned are not to be found, then you could increase the size of the hole to 40 mm ($1\frac{1}{2}$ in) in diameter. This will make the entrance wide enough for the house sparrow to use. Although these birds will almost certainly have plenty of natural nesting sites – too many, perhaps, some people would claim – you might consider that it is better to have a bird of this type nesting in your garden than no bird at all. In some areas, birds might try to enlarge the original hole, and to prevent this it is possible to fix a tin plate, preferably inside the box. The problem here is likely to be the cutting of the metal to the actual shape. In addition, care must be taken that there are

Nest of a goldfinch

no sharp edges on which the bird might cut itself.

Siting the nest-box

Once you have your nest-box you will want to know when, where and how to fix it. The box is secured to either a tree or a post by the piece of wood at the top. However, it should not be attached directly to the structure. If it is, water will almost certainly trickle down the back and cause the box to become damp and eventually to rot. A piece of wood should be screwed to the tree or post, and the box attached to this. It must also be borne in mind that the top of the box should be slightly angled. As well as preventing rain from entering this will also stop direct sunlight from disturbing the offspring.

Traditionally nest-boxes have been fixed to trees, but this is not essential. In fact the RSPB has even suggested that where suitable secluded sites are not available a box can be attached to a post in the open. Position in terms of the sun, wind and rain is very important. Prevailing winds will obviously cool down the box considerably, and may also drive the rain into the hole. In selecting sheltered sites it should be remembered that the nest-box needs to be visible in the first instance and that birds which choose to nest there will also need to be able to have a clear flight path to and from the hole. Because birds do not like competition, one should endeavour to place the next-boxes apart from each other. Although those birds which have their own territories will not tolerate intrusion from their own kind, they do seem to live peaceably with other species.

When fixing nest-boxes to trees, care should be taken. There are some species like the oak, which are particularly durable, and by screwing a nest-box to the trunk little harm is likely to be done. However, other trees are more susceptible and in this case wire should be used. Hooks can be screwed into the box and the wire put through these and around the tree. A temporary fixing might be best in the early stages. You might decide that the situation is not particularly suitable for your box and you may want to move it to a new and hopefully more advantageous site. You will also want to remove it at the end of each

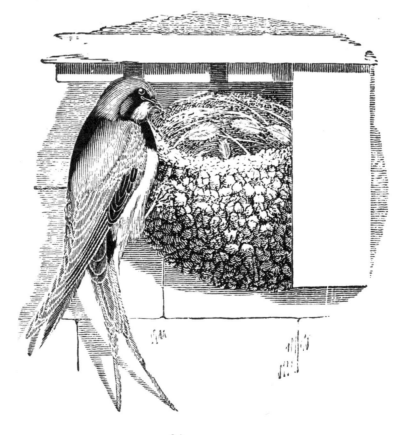

Swallow and nest

34

season so that you can give it a good clean out.

Although the birds which use the nest-box will not breed until the spring, the boxes should be put in position in the previous autumn or winter. Since many birds will start to look for nesting sites a while before they actually mate and lay their eggs, it's a good idea to have your bird-boxes in position so that they become part of the scene, and are not new and strange. Furthermore, they could act as a valuable winter roosting quarter for birds in harsh weather. Wrens in particular will huddle together inside suitable boxes and several records of these birds during the harsh winter of 1962–63, and the cold snap of 1978 confirm this.

Although we have pointed out that the non-hole-nesting species can usually find suitable sites in and around many gardens, there are nest-boxes which can be provided for them. It is quite permissible to produce the same standard design as the hole nest-box. In place of the hole in the front, a piece of wood is screwed on, but this is only about half the depth of the front. This leaves plenty of room for the birds to get in. The alternative to this design is simply an open tray. Any artificial nest should be placed in the fork of a tree. It is not necessary to provide as many of this type of box as those for hole-nesting species. If you have a smaller garden you may only have room for the hole-nester's type. Don't be surprised if some tit species, like the blue tit, starts to attack the hole. They seem to prefer a nesting-hole which they have 'improved', a task which they perform avidly when selecting natural sites, and so it is not surprising that they try the same sort of approach with an artificial box. Of course the first reaction is to assume that the hole is not of the correct size. However, if you have made it 28 mm (1⅛ in) as suggested, don't do anything to increase the diameter. If you do, the chances are that you will make way for the starling and lose your blue tits. Starlings will in any case often scare off a blue tit even after it has decided to take up residence. There is not much that you can do, because the chances are that you will be out or at least busy, when the starlings take up the cudgel. Some people have suggested that if the box is placed conveniently, you could have a remote control flag, so that when you pulled the string the flag would move about. If a blue tit has decided to take up residence, and is then threatened by an over-zealous starling, the temptation is to move the nesting-box to protect it from the intruder.

Wren's nest

Don't take this action, because if you can't deter the starling with the nest-box in its original position, you will undoubtedly lose the battle and the blue tit anyway.

Nesting materials

Although we have suggested that you should not interfere with your nesting bird, there are ways of helping it. During the nest building operation many species will eagerly snatch up anything offered to them for feathering their nests. You could fill small mesh bags, like those which contained peanuts, with odd bits of wool and cotton, and if you have found sheeps' wool locally this will also be useful. Even though the birds might not actually nest in your garden, many will visit it and discover the nesting materials, which they will carry off to their own nesting sites. It is not a bad idea to have two bags in your garden: thread some string through one, and hang it from a branch, because some birds prefer to collect their material from above. Leave the second bag on the ground for those birds which like to find their materials here. This bag needs to be fixed in position, otherwise it will blow away. A piece of wire from one of those coathangers which dry-cleaners present to their customers will hold the bag to the ground.

Observing your nest-box

If a nest-box becomes occupied you will undoubtedly want to peep. In the wild the chances are that from time to time the bird will be disturbed. This is a fact of life, and she will usually make light of the intrusions, carrying on with the job which she has to do. You obviously want to know what is going on in your nest-box and so the occasional peep is permitted. You do not have to frighten the birds away and *it is very important that you exercise great care.* With patience you will soon be able to discover when the bird is at the nest and when she is away. Once the eggs have been laid it is imperative that the bird keeps them warm so don't frighten her off the nest. With most species she, or he, depending on who is doing the sitting, will leave from time to time. Once the eggs have hatched, you will usually see the two parents bringing a continuous supply of food for the hungry young. There is no hard and fast rule about how often to view, although every three or four days is probably about right. You should do your viewing when the parents are away from the young. Particular care should be exercised just before the young are ready to leave the safety and seclusion of the place which has been their home. If you disturb them before they are ready it is likely that they will be frightened, which makes them panic. If this happens, they will leave before the time is ripe, so that they fall prey to predators out in search of food, or perhaps die from starvation.

Cleaning the nest-box

Once your birds are fully fledged and flown, you probably want to clean out their nest-boxes. Many do become rather foul, and they will certainly harbour parasites and other invertebrates, which can feed on any remains in the nest. If a nest-box is suitably positioned mice may take up residence; a box belonging to a friend actually had bees in it, which kept the birds away! Many people are surprised, and perhaps not a little dismayed, to find addled eggs or dead young in the nest box. It is quite natural and nothing to be alarmed at. It could be that one of the brood was weak, or had a deformity, so that if it had survived at first, it would almost certainly have died very quickly. Large numbers of eggs are laid to counteract the mortality rate and natural wastage which is much greater than for man. If you are handling material from the nest, use gloves and do it carefully. You might want to take the nest to pieces to see how resourceful the bird has been, but it is a good idea to fumigate the nest first, so that the chances of disease are minimized.

Unusual Nesting Sites

You might have seen pictures of some of the strange places where birds have nested. The underneath of car wings and bonnets are not unusual sites. You could provide your own unusual items. An old teapot, provided that it has a large enough hole where the lid was, will often become the residence of various species. It needs to be tucked into any ivy-clad hedge or fence so that, hopefully, it cannot be seen or attacked by cats. If you are using a teapot or kettle, apart from removing the lid you should remember to make sure that the spout is in the *downward* position, so that it does not collect rainwater.

Nest-boxes from logs

Another unusual nest-box can be made from a log. The diameter of the particular log used is not of any great importance, but if it is less than about 15 cm (6 in) in diameter it might be difficult to chop. If you find a suitable piece of log you can cut it into lengths of about 25 cm (10 in). You will need to cut off a piece at the top and a piece at the bottom for the lid and base: these have to remain intact. Measure 25 mm (1 in) from the top and the same distance from the bottom, and then saw these pieces off. You can now proceed with the rest of the operation. Mark an X across the top of the log, trying to ensure that all four sections are equal in size. Use a small axe and mallet, and split the log along the lines into four pieces. Mark the outside 1, 2, 3 and 4 (working clockwise) so that you will know how to join them later. Again using the axe and mallet, split off the middle section of each piece of log, to leave about a 25 mm (1 in) wall: this will be enough to ensure that the pieces can be re-joined. This can now be done by gluing the sections together, although you might find that staples are quicker. After joining the four sections, fix the bottom on. Glue is to be preferred, but nails can be used. If using either nails or staples ensure that they do not stick out inside the box. Glue should be of the waterproof type: it would be disastrous if the box fell apart when the hen bird was laying her eggs or tending her young! To fix the roof

Redstarts find an unusual nesting site in a teapot

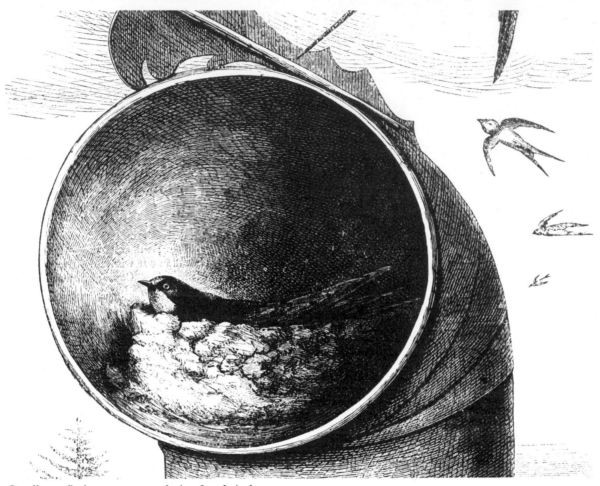

Swallows find an unexpected site for their home

either screw it on or use a hinge. If using the latter a hook will be necessary to secure the lid. Hinges made from webbing are better here because of the shape of the box. Another point which should not be overlooked is that when sited the box should slope so that water will run off. It might be an idea to have logs sawn at an angle, so that the box will have a natural, sloping roof. When fixing to the tree you can use wire, but make sure that there is a batten between the box and the tree, so that water can drain down the back.

A special nest-box

An unusual nest-box, close enough for you to view, can be attached to the outside of one of your windows. The only difference between this box and a conventional one is that it does not have a back: your window pane serves this purpose. You will need to cut a piece of opaque card or other material which

you can attach to the inside of the window. Once your box becomes occupied you will be able to peep at the activities going on inside it. As with an ordinary box, you will need to exercise caution, and ensure that you don't disturb the birds, otherwise you might have a disaster on your hands. We suggested that other nest-boxes should be put up in autumn/winter, and this rule also holds good for the window box. Of course you cannot guarantee that a bird will actually take up your offer of hospitality, but you might increase the chance by placing food in the vicinity during the winter months.

Nests in walls

If you can provide a suitable habitat in a wall birds might take up residence. When building your own wall leave out half a brick to provide a useful nesting site. There are other birds you can encourage, which, though they will almost certainly never use any kind of nest-box, will nevertheless make use of extra spaces, and those likely to occur in the garden are the swallow and the swift. If you have a suitable outbuilding, like a garage, you could make sure that there are holes in the walls or doors so that swallows can get in. Provided that there are suitable rafters they will build there. Of course you will probably find that your car will become covered with droppings, and it might not be a bad idea to leave it outside during the breeding season! Swallows will often return to the same nesting sites year after year. I remember an outside privy in the Norfolk village where I lived as a lad. For several years swallows always nested in the rafters, making their entry at the top of the ill-fitting door. This is not so long ago, and there are prob-

ably still plenty of these outmoded outhouses which will be occupied by these birds if there is a space through which they can gain an entry.

In such a short space we have only been able to deal with common aspects of nesting sites and boxes. It may be that you live in wild countryside, and have kestrels coming to your garden – actually you don't have to live in that wild countryside, because these birds of prey have bred in London! In this case you might decide that you want to provide suitable nesting boxes for them. For information about the type of box suitable for more unusual birds you can do no better than obtain the excellent publication *Nest-Boxes*, which is available from the British Trust for Ornithology and details of which are given at the end of this section.

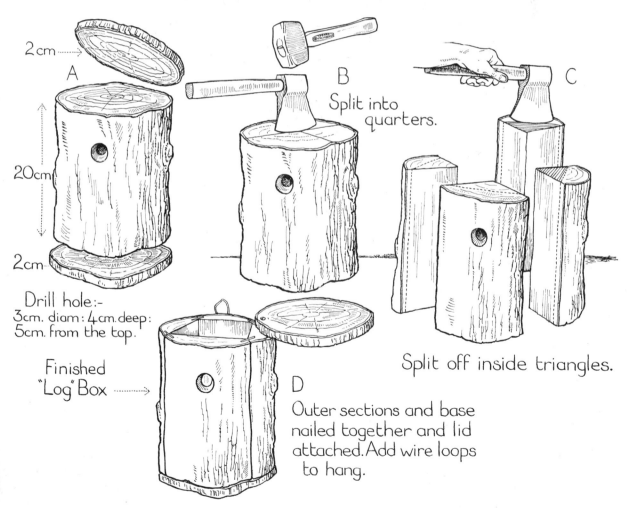

2 cm

A

20 cm

2 cm

Drill hole:-
3 cm. diam: 4 cm. deep:
5 cm. from the top.

Finished
"Log" Box ·······>

B
Split into
quarters.

C
Split off inside triangles.

D
Outer sections and base
nailed together and lid
attached. Add wire loops
to hang.

How to make a nest-box from a log, as described on opposite page

A young cuckoo is fed by a dunnock

Coping With Bird Casualties

It is quite possible that sooner or later you will have to deal with casualties, and under this heading we have included birds which appear to be deserted, have fallen from their nests, or have lost a parent. Quite often the good intentions are totally unnecessary, because the young birds are not really waifs and strays at all. In fact it is highly probable that instead of doing a service to the birds, you will be doing them more harm than good. You will need to differentiate between a destitute youngster and one which has simply lost its mum temporarily.

Casualties, as opposed to deserted young, are another matter. Young children are quite distressed when they find a bird with a broken wing or a broken leg. Sometimes the signs of injury are not so obvious. If you feel that you cannot cope at all the Royal Society for the Prevention of Cruelty to Animals (RSPCA) or the People's Dispensary for Sick Animals (PDSA) should be notified. You will be able to find their telephone numbers in the local directory. The PDSA has a number of animal treatment centres situated in many towns, and if you are able to get the bird there they will certainly do their best to deal with it. A telephone call before setting off is advisable to check opening times and save needless journeys. It is likely that if the bird can be treated, you will be offered it back with details of the way in which it can be helped back to full mobility and subsequent freedom. In many areas there are also local animal welfare groups, which might be able to deal with the casualty.

Stunned or shocked birds

Birds have a nasty habit of flying into our large windows. It seems incredible that so small a form should make so great a noise as it hits the window, and perhaps what is even more remarkable is that, although the bird is stunned and 'knocked out', there is often no other damage, and once it has recovered it is usually able to fly away. The greatest danger, and almost certainly the cause of most deaths, is not, as would probably be expected, any form of injury, but the shock associated with the mishap. The bird needs similar treatment to a human being when suffering from shock: warmth and quiet are essential requisites. If you find a bird which has been stunned handle it carefully when you pick it up. You should always have your 'hospital' box ready. All that you will need is a shoe box which has some soft material in the bottom and some ventilation holes punched in the lid. The casualty should be placed inside the box (with the lid in place) and the box kept in a quiet corner. Once the bird comes round you will soon hear it fluttering as it tries to escape from its 'prison'. Rather than handle the bird, which might give it another shock, the best course of action is to take the box outside into a quiet place and remove the lid. Unless the bird is injured, it will quickly seek its freedom.

If you don't have a hospital box, you can nurse the bird yourself. Either gently wrap it in your jumper, ensuring that it can breathe, or cradle it in your cupped hands. How long the bird will take to regain consciousness will depend on the force of the initial impact. Once the bird begins to come to, make sure that it is not frightened and take it calmly into the garden so that it can fly away. On some occasions either the actual impact or the shock will have killed the bird. Sometimes it may have undetectable injuries, in which case you may wish to make sure it is humanely destroyed as soon as possible.

Broken bones

Dealing with casualties other than stunned birds demands patience, rather than a great deal of knowledge of the bird's anatomy, although the latter does help. Depending on the bird's injuries, the amount

Cuckoo being fed by wrens

of time which you need to spend on your casualty could be considerable, and you should bear this in mind before taking a decision on your course of action. Children tend to become very attached to birds which are kept in captivity, and since your bird may be with you for quite a while you must make it quite clear that it has to be returned to the wild as soon as it has recovered.

It is difficult to discover whether a bird has internal injuries, whereas a broken leg is usually very evident. If you decide to pick up an injured bird to examine it, you must do so carefully, and it is advisable to have gloves on. This is not because the bird is likely to peck you, although this is often a natural reaction, but because some birds carry diseases, which you might catch. With smaller birds, gloves make handling difficult, so it is important to remember to wash your hands well after examining the bird. If you decide that the bird is ill, rather than injured, then you could consult a veterinary surgeon, although consultation and any subsequent treatment could prove expensive and most sick birds do not recover even if they are treated professionally.

There are ways of mending simple broken bones, especially where the injury is to the lower leg bone –

Skylark and hungry young

the tarsus. All that you will need is some sellotape and two matchsticks, or blunt cocktail sticks. You will need a cage as your hospital ward. You will have decided that the leg was broken because it was 'bent'. Before putting on your matchstick splint, straighten the leg carefully, using as little pressure as possible. With the help of another person place the sticks on opposite sides of the leg, and bind the whole with sellotape. The bird can now be placed in the cage. If you don't know what bird it is use a book to identify it. This is important since you will need to feed it correctly while it is in your care. You can find details of its natural food in a book like the *AA Book Of British Birds*. For seed eaters you will be able to obtain seed mixtures and feed these in the same way as you would on the bird table. Insectivorous birds can usually be fed on mealworms which you can purchase, or they can be offered earthworms which you can dig up in the garden. Don't forget that your casualties will also need water, which must be changed regularly. You might also find that as your bird recovers it needs a container such as a shallow dish in which to take a bath.

Provided that the bird doesn't have any other injuries, and you have set the leg carefully, it should recover within a few weeks. It is better to keep the bird in captivity for too long rather than set it free before the bone is healed. The perches in most cages are unsuitable for wild birds and if possible should be replaced by small branches, which will give the bird a natural foothold.

Injuries from cats

There is a natural order of things, even in the garden, and the fact that most cats have a natural instinct to kill birds will undoubtedly lead to injury at the best and death at the worst in many cases. Once a cat has got hold of a bird its claws will have probably gone quite deep into the flesh. Death is likely to result either from shock or from injuries received, or both. If it appears that the wound is very severe then the best course of action is to have the bird humanely destroyed. Without causing the creature any suffering you can kill it yourself, as long as it is only a small bird, by firmly pressing the neck with your fingers. Death will occur quickly. With a large bird the problem is more difficult, and it is safer to take it to a vet or animal welfare organization so that it can be humanely destroyed.

Wren parents and their nest

Blackbirds

fully bringing it up is probably remote unless you have a great deal of time and patience to devote to it. It is better to humanely destroy the bird than to start to care for it, only to find that the demands are too great, and the only course of action is abandonment.

As in the natural state, nestlings have to be fed regularly, and warmth is essential if the bird is to come through its ordeal. Your hospital box, mentioned earlier, will be suitable for your orphan, although you might need to transfer it to a deeper one as the bird grows and tries to escape, as it will inevitably do. Ensure that there is a piece of warm cloth on the top of the material in the box – flannel is ideal for this purpose. The box, like the nest, needs to be kept warm at about 16°C (61°F), and in a draught-free atmosphere.

For most of the time the nest is generally quiet and dark, and your artificial home should provide a similar environment. It is a good idea to have the lid of the box fastened with elastic bands, which will allow ease of opening when feeding, but which will deter people from continually peeping and disturbing the bird.

Feeding is a problem, particularly at first. When you look at birds in the nest they appear to have ever-open beaks, and yet when they are being hand-

Dealing with orphans

In the case of birds which appear to be orphaned it is necessary to make sure that the youngster is not just lost. Although one's instinct is to feel sorry for what appears to be a helpless, forlorn nestling or fledgling, don't let your heart rule your head. In most cases the parents will be close by, probably seeking out food for the young bird. Children are tempted to pick up a fledgling and take it home. Try to dissuade them from doing this, but if you are presented with one, take it back as quickly as possible to the place where it was found, and the likely outcome will be that an anxious parent will soon deal with the 'lost' soul.

If you do feel that a bird has been abandoned by the parents you should first try to make sure that it is left where you first noticed it. Leave it here for two hours or so and you will usually find that the parents have discovered their offending offspring, and it will be safely returned to the fold. Of course there are occasions when this does not happen, and you can be fairly certain that the bird has been orphaned if it is not found within two hours. The chance of success-

Sagacious sparrows

43

fed they are generally not very cooperative. A small piece of bread, preferably carefully rolled into a small ball not much bigger than a matchstick head, should be offered on the clean end of a matchstick. If the mouth is open you won't have any problems and can gently push the food into the gullet. If the bird is uncooperative you will have to gently force the beak open. Unless the orphan is ill it will soon manage to open its mouth when food is offered.

The task of feeding the bird will be time-consuming. You will need to be up at the crack of dawn to give it its first meal, and then you must continue feeding it every hour until sunset. As the bread can only be fed for one or two meals, you will need to prepare a special mixture, depending on whether the bird is a seed-eater or an insect-eater. The seed-eaters require crumbled biscuit, to which have been added a few drops of cod-liver oil and some egg. The insect-eater will need the same biscuit meal – oatmeal or wheatmeal biscuits are particularly nutritious – to which has been added a small amount of the yolk taken from a hard-boiled egg. This should all be mixed to a paste with water. Make small pellets about the size of a matchstick head, and offer these in the way suggested earlier. The mixture should only be made up

The swift

in small amounts each day, so that it is fresh.

A bird at this age needs a few drops of water each day. You can give this from a dropper or from the end of your finger.

How do you know when the bird has taken enough? It may try to stop being fed before it is full. The food collects in the crop at the base of the throat, and you should see this swell. When it appears fully extended you should stop even if the bird begs for more. Generally, however, most young

A sparrow feeds its young

birds will know when they they have taken enough nourishment.

Once fully feathered the bird needs to learn to feed on its own. Always leave some food in the box – in a container and not loose, so that it does not become contaminated with the bird's droppings – and the bird will be able to help itself. You should gradually ease off your hand-feeding to encourage the bird to take its own food. With some birds it may be easy: with others more difficult.

There are important daily jobs which must be attended to, and the fledgling's home must be kept clean. Fortunately droppings are expelled in a small jelly-like bag, which can easily be removed with forceps. Failure to keep the nest-box clean could result in the bird's death. Check the bird as well, as in the wild it will usually be tended by the parents. The feathers at the anus may become dirty, and to avoid infection should be cleaned with a cotton bud, dipped in warm water. It is essential to keep the 'nest' clean. Dirty nest materials should be removed regularly. The cloth can be soaked in disinfectant, washed and dried to be used again, or destroyed.

The most difficult part may be acclimatizing the

bird to the wild state when it no longer needs your care. Once the feathers have appeared and it seems ready for flight you should let it loose *inside* the house. Make sure that the room is darkened, because the light from a window will lure the bird in that direction, and it could injure itself as it hits the glass. Once flying, it should be taken out into the garden, so that it can obtain its own food and fend for itself.

An interesting and important point which is seldom taken into consideration when dealing with allegedly orphaned species is that many do leave the nest before they are ready to lead independent lives. The birds in the wild countryside will at this stage still be tended and cared for by the parents, generally without inquisitive man interfering with nature's course. This is likely to happen in the garden too, and *it is imperative that you determine the bird has no parents before you attempt to rescue it.*

Further Useful Information

Organizations and Societies
BTO (British Trust for Ornithology), Beech Grove, Tring, Herts, HP23 5NR

The BTO carries out various research investigations, including the very important Common Bird Census, which gives valuable information about the status of Britain's birds. It also publishes *BTO News* and *Bird Study*. Further information about the Trust, and the work which members can do is available.

RSPB (Royal Society for the Protection of Birds), The Lodge, Sandy, Beds, SG19 2DL

This is the foremost British bird 'conservation' society, and welcomes new members who have an interest in its work. It publishes *Birds* four times a year. Further information available.

Periodicals and Magazines
Birds, RSPB, The Lodge, Sandy, Beds, SG19 2DL, is available to members
Birds International, ICBP Circulation Dept, Edward Wright Ltd, 5-7, Palfrey Place, London SW8
Birds & Country Magazine, 79 Surbiton Hill Park, Surbiton, Surrey
    Published quarterly
British Birds, Macmillan Journals Ltd, Brunel Road, Basingstoke, Hants, RG21 2XS
    Published monthly on subscription

Books, Leaflets, etc
Armstrong, E. A., *The Wren*, Collins
Barrington, R., *The Bird Gardener's Book*, Wolfe
Book of British Birds, Readers Digest/AA
Bruun, B. & Singer, A., *Birds of Britain & Europe*, Hamlyn
Burton, P. & Hayman, P., *The Birdlife of Britain*, Mitchell Beazley
Campbell, P., *The Oxford Book of Birds*, OUP
Campbell, W. D., *Birds of Town and Village*, Hamlyn
Christie, Cliff, *And Then They Fly Away*, Constable
Cowan, T. A., & Barnes, J. A., *Birds of the British Isles and Their Eggs*, Warne
Evans, G., *The Observer's Book of Birds' Eggs*, Warne
Fisher, J. & Flegg, J., *Watching Birds*, Poyser
Fitter, R. S. R., *Bird Watching*, Collins
Flegg, J., *Binoculars, Telescopes and Cameras for the Bird Watcher*, BTO
Flegg, J., *Nestboxes*, BTO
Flegg, J., *Discovering Bird Watching*, Shire
Harrison, C., *Nests, Eggs and Nestlings of British and European Birds*, Collins
Harrison, R., *A Beginner's Guide to Bird Watching*, Pelham

The nuthatch

The redstart

Heinzel, H., Fitter, R. S. R. & Parslow, J., *Birds of Britain and Europe*, Collins
Hicken, N. E., *Bird Nestboxing*, Stanley Paul
Hollom, P. A. D., *Popular Handbook of British Birds*, Witherby
How to Begin the Study of Birds Around the House, BNA
How to Begin the Study of Birds, BNA
Jones, R., *Birds in Our Gardens*, Jarrold
Jones, R., *Birds of the Hedgerows and Commons*, Jarrold
Knight, Maxwell, *Bird Gardening*, RKP
Perrins, C., *Birds* (Countryside Books) Collins
Peterson, R., Mountford, C. & Hollom, P. A. D., *Birds of Britain and Europe*, Collins
Saunders, D., *RSPB Guide to British Birds*, Hamlyn
Soper, T., *New Bird Table Book in Colour*, David & Charles
Soper, T., *Everyday Birds*, David & Charles
Summer-Smith, J. D., *The House Sparrow*, Collins
The Birds in Your Garden, RSPB
Vere Benson, S. *The Observer's Book of Birds*, Warne
Wild Birds and The Law, RSPB

Equipment and Supplies

Adastral Products, 517 Portswood Road, Southampton:
   Nesting-boxes
Ambig Ltd, 3 Baronsmead Road, London SW13 9RR:
   Seed dispenser – the Dinabird
Coombs, E. W. Ltd, 25 Frindsbury Road Strood, Kent:
   Various bird foods, including mealworms and seed mixtures

Dendy, R., 2 Aultone Yard, Aultone Way, Carshalton, Surrey, SM5 2LH:
   Rikden birdbaths and small bird feeders
Forest Drums Ltd, P.O. Box 60, Southampton, SO9 7ED:
   Water butts
Greenrigg (Birdcraft Products), 1 Stanley Road North, Rainham, RM13 8AX:
   Various feeders, *etc*
Haith, John E. Ltd, Park Street, Cleethorpes, Lincs:
   Bird seeds – wide variety, and cheap if bought in bulk
Kings House, Nadderwater, Exeter, EX4 2LD:
   Window-sill bird feeder – sae for illustrated leaflet
MGR, 12 Haycroft, Bishop's Stortford, Herts, CM23 5JL:
   Merry Go Round Bird Feeder, for tits. Ready to assemble
RSPB, The Lodge, Sandy, Beds, SG19 2DL:
   Produces a wide range of garden bird equipment. Their illustrated catalogue is available on request to non-members – sae please
Wood, Jamie, Ltd, Cross Street, Polegate, Sussex
   Wide variety of feeding trays, bird tables, nesting-boxes, *etc*. Also hides

Miscellaneous

Bird Bookshop, 21 Regent Terrace, Edinburgh, EH7 5BT:
   Natural history books, especially those dealing with birds

The swallow

Swallows preparing to migrate

British Garden Birds – two records and book, by Peter Conder
Evans, David, Fine Bird Books, The White Cottage, Pitt, Winchester, Hants:
  Bird books bought and sold
Humane Education Centre, Avenue Lodge, Bounds Green Road, N22 4EU
  Birds in Your Garden – a record by Percy Edwards
Kirby, John, 10 Wycherley Avenue, Middlesbrough, Cleveland, TS5 5HH:
  Bird song on cassette – No 6: *Garden Birds*. Send

sae for details
Garden and Park Birds, Shell Nature Record, from Wildlife, 243 King's Road, London SW3 5EA
RSPB, Sales Dept, The Lodge, Sandy, Beds, SG19 2DL:
  Has a wide variety of material for sale, including records, posters, wallcharts, *etc* – all in their illustrated catalogue
Wild Life Sound Tracks – No 6: *Common or Garden Birds, RSPB, The Lodge, as above*

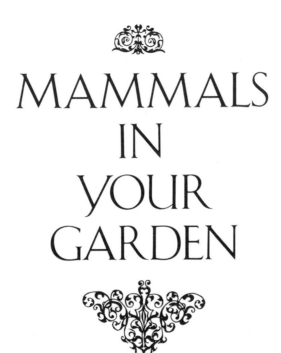

MAMMALS IN YOUR GARDEN

The illustration shows (clockwise, from bottom):
hedgehog eating a worm, bats against the house, fox
getting ready to raid the dustbin, hazel nuts (the broken
ones eaten by a squirrel), long-tailed field mouse

Although you are probably not aware of it your garden is likely to be a regular haunt, perhaps even residential quarters, for all sorts of mammals, even though many mammals are secretive and nocturnal, and you will probably not have seen them.

Mammals On Your Patch

If you haven't actually seen mammals you have probably spotted signs. It might have been the cigar-shaped droppings of a sniffing hedgehog, or the identifiable footprints of a marauding fox. If foxes and hedgehogs are visitors, the likelihood of other nocturnal visitors cannot be ruled out. What sort of night-time 'intruders' are likely to come to your garden? There are mice, rats, voles, hedgehogs, and if you are lucky, perhaps old brock himself. Many of the mammals of the so-called country-side have been visitors to the urban-dweller's garden for perhaps hundreds of years. Many of the rodents have lived in close association with man, taking advantage of the food and hospitality which he has to offer, for although man terms much of what he throws away 'waste', many animals find this a valuable source of nourishment. As man has depleted the countryside and increased his waste, new species are now taking advantage of the urban situation. Foxes are known to be regular visitors to suburban gardens, and it seems that some have managed to penetrate far into the centre of many of our towns and cities.

So far we have mentioned the mammals which make their way on foot; we mustn't forget the only true flying variety – the various pecies of bat. As most people will know, church towers are a favourite haunt of bats, and since these are often in the centre of urban populations, bats could be expected in any garden. But churches are not their only

hang-ups, if you'll pardon the pun. They will be found in derelict buildings, and quite frequently in the lofts of many houses.

The Mouse Family

Mammals have minds of their own and can roam at will. However, there are certain gardens which will provide an environment which at least might prove hospitable for visiting mammals. Perhaps the most obvious are the mice. The two species of mice which are likely to be unseen residents are the long-tailed field (or wood) mouse and the harvest mouse. Of these animals, the field mouse is more nocturnal

Long-tailed field mouse

than the other species. The harvest mouse has definitely decreased in numbers, although no accurate figures are available.

The long-tailed field or wood mouse

Areas of rough grass will provide a suitable habitat for the field mouse, which makes a nest in a hole in the ground. This might be sited amongst the exposed roots of a tree or in rough grass. Once a spot has been selected, grass and dead leaves are used to line the nest. This species is fond of many of the berries and fruits which may be found in your patch. Hips from the rose bush and haws from the hawthorn hedge will prove a magnetic attraction, and so the mouse is a common visitor to some gardens. Although quite shy when compared to the harvest mouse, the field mouse has taken advantage of man, and will take up its position in both the house, if it

The mole

gets the chance, and in outbuildings, like sheds and garages. It will, however, fight shy of these areas if the house mouse is already in residence. The garden hedge is also a favourite haunt, because here the animal has relative security. Where there is a good supply of leaf litter it will make numerous runs.

During the summer the field mouse is a prolific breeder. Several litters are usual in a few short months and, as with the harvest mouse, there may be up to nine youngsters at a time. With a plentiful supply of food in the late summer and early autumn the field mouse will collect and carry food to its nest to be stored for the forthcoming winter period. These are the creatures which come out from their homes to raid the garden. They are usually guilty of removing beans and peas before they actually germinate. Thus if you have, and want to encourage these creatures to your garden, it is worth waiting until around the end of April before planting these crops. They seem to survive and grow at this time, probably because there is new growth appearing in the countryside, which provides an alternative source of food for the mammals.

The harvest mouse

If you live in the part of the country where the harvest mouse is still relatively common, and you

Harvest mouse

have an area of tall, rough grass, then this mammal might build its nest in your garden. During the summer, the mouse will often move in from the surrounding countryside to take up a temporary abode in your habitat. Its nest is made in rough grass. Here it will use other dead stems which it twines and intertwines around the upright stems. The nest, supported by the stems of tall grasses, will serve it well. From here it will make excursions in search of a continuous supply of food. Agility is the creature's forte, as it moves among the stems of tall grasses and corn. Most of the harvest mouse's life is spent above ground and as a breeder it is prolific.

Short-tailed field mice

Common doormouse

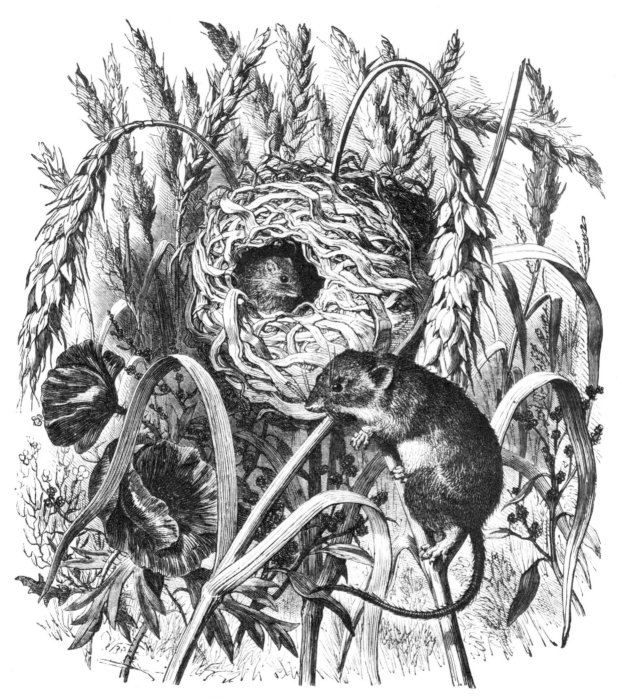

THE NEST OF THE HARVEST MICE.

The nest of the harvest mouse

52

There may be up to four litters a year, and with up to nine young per birth, the numbers could quickly increase in any area, though normally this does not happen, as it has many predators. Having produced its summer nest the harvest mouse usually stays in residence well into the autumn, and many mice have been discovered still occupying their homes well into November.

The Hedgehog

Although, like many other mammals, the hedgehog is nocturnal, of all the animals found in, and associ-ated with the garden, this species is considered *the* garden animal. Yet in spite of its fame, and the affection with which it is usually regarded, how many people have actually seen it? Although relatively easy to identify, the cigar-shaped droppings which it leaves behind are difficult to discover unless the hedghog has left its trail as it crossed the lawn. Most will be lost either in the hedge bottom, or amongst the shrubs in the border. Although the animal has a distinctive track, as you can see on page 63, this is only semi-permanent, and will usually disappear quite quickly.

The hedgehog has various names in different parts of the country, and perhaps it isn't surprising that, with its sniffing, snorting noises, it has become known in many places as the hedgepig. It is not uncommon for the animal to let out a pig-like squeal which gives a further reason for this name. One of its other names is urchin, and, according to a rhyme, it was once possible to 'tell' the weather by the hedgehog:

Observe which way the hedgehog builds her nest,
To front the north or south, or east or west;
For if 'tis true what common people say,
The wind will blow the quite contrary way.
If by some secret art the hedgehog know
So long before which way the winds will blow,
She has an art which many a person lacks
That thinks himself fit to make almanacks.

If you have been out in the garden at night, you may have heard an unexpected and curious noise coming from your herbaceous border or hedge bottom. The sounds, a mixture of snorts and snuffles and grunts, are made as the hedgehog makes its regular nightly pilgrimage for food.

Courtship and mating

If you are lucky enough to have hedgehogs in your garden you might witness the mating ceremony, which is performed with great gusto. Actually you are hardly likely not to hear it if the creatures decide to act out their dramatic courtship ritual close to your window! Most people generally have a picture of hedgehogs as docile creatures, despite the fact that they do have quite a turn of speed when they need to make a quick getaway! Mating can take place any time from May to August. The hedgehogs are quite active during their mating preamble. They will chase each other, and actually leap into the air from time to time. But movement is not the only part of the display; the squeals of delight, punctuated by loud grunts and snorts, can often be heard over quite an area. The 'games' over, the hedgehogs mate, and about thirty days later a litter will be born. If the pig, the female, is a resident in your garden, and there is a quiet area for her nest and confinement, she may well give birth to her young in your garden, right under your very nose, and you will have not one or two, but a family to keep down pests!

Should suitable hibernatory quarters be available the hedgehogs will usually take continued advantage of your hospitality. The temperature is a controlling factor in determining the point at which the hedgehogs call it a day, and retire for the winter. Actually, they are not true hibernators, for they will usually wake up at some stage during the winter months. Generally, the temperature is low enough towards the end of October to signal that it is time to go out of circulation for a few months. However, if the winter turns out to be a mild one, these mammals are likely to be around later in the year, and if these conditions prevail it is not unusual to see them in December and January.

Encouraging the hedgehog

The hedgehog is to be encouraged in the garden because of its usefulness as an almost universal pest controller. The search among the plants in the herbaceous border and in the hedge bottom goes on for quite a while every evening, until the hedgehog has satisfied its hunger. Hidden in among the plants which make up the border, and safely concealed in the dead leaves of the hedge bottom, the urchin will find an almost endless supply of insects and their larvae, as well as slugs. The same pests will be lurking in your compost heap, if you have one, and the hedgehog will take advantage here too of what seems to be an unlimited supply of food.

While one doesn't want to 'tame' wild creatures, it is useful to encourage the hedgehog, and the mammal is very partial to milk. If you think that a hedgehog is on the beat in your patch, you could leave out some milk for it. As the sun goes down the hedghog will stir itself from its diurnal slumbers, ready to go out on its nocturnal jaunt. If you have some form of hide or observation point in your garden (see page 61) you can use this as a base for watching your hedgehogs. If you can actually watch a hedgehog lapping milk you will almost certainly be amazed. It laps it up as fast as a cat. But the cat doesn't usually manage to get its snout into the milk; many hedgehogs do, and if this happens the animal will spend a little while performing a grooming operation before making off into the night.

Once you have a hedgehog in the garden there are ways of 'keeping' it there. During the late spring, when the animal emerges from its winter sleep, through to autumn, the large garden will usually provide enough food for the hedgehog to make regular visits. If you want him to become resident, then you will have to take some precautions. Hedgehogs are fond of hedges, and if your garden has a length of hedging, then you will at least be partly home and dry. Wire netting can be sunk into the ground, to a depth of about 20 cm (8 in). One-metre high wire netting is suitable for this purpose. If it is let into the ground on the outside of your boundary this will allow the hedgehog to have access to your hedge, while preventing it from leaving your garden. However, hedgehogs do have the ability to climb, although metre-high wire netting will usually act as a deterrent. Young hedgehogs are more adventurous than their elders. Those which do decide to climb seem to do so with amazing agility. You might think that once the top of the obstacle has been reached, the hedgehog will be stranded, but it will simply roll itself into a ball and then set off, gathering momentum as it moves towards the ground!

The spines, although giving the impression of being totally rigid, do have some flexibility in the area where they join the body. The hedgehog's spines are of great importance as a means of protection – against most species anyway. But such a system does make it difficult for the animal to keep itself hygienic. Most of the hedgehog's body is like a brush, but as he has great difficulty in keeping the bristles clean, he is permanently infested with fleas – so beware!

Making a house for your hedgehog

As a result of its research the Henry Doubleday Research Association has produced a design for a hedgehog house, which is illustrated. One of its characteristics is that it can be produced quite cheaply and effectively by even the most clumsy handyman as long as he can manage to wield a hammer, saw and brace and bit.

You will need the following wood:
  3 m (10 ft) pieces of 5 x 2.5 cm (2 x 1 in)
  4.5 m (15 ft) lengths of 10 cm x 18 mm (4 x $\frac{3}{4}$ in) planks
  38 cm (15 in) lengths of 10 cm $\times$ 18 mm (4 $\times$ $\frac{3}{4}$ in)
The wood can be purchased either rough-sawn or ready-planed. The former is obviously cheaper. You may also be able to obtain some of the wood from a local demolition site. One word of advice: although you might be tempted to creosote the house to protect it from the weather, if you do so you will certainly not attract your hedgehogs. They don't seem to like even the faintest smell of either creosote or paint. For preference untreated wood should be used, although if you do need to use something then the wood should be treated with clear Cuprinol.

The entrance tunnel is the first part to be made, and in it four pieces of 10 cm x 18 mm (4 x $\frac{3}{4}$ in) are fitted so that they make a run which is 10 cm (4 in) high and 7.5 cm (3 in) wide. A large hedghog will be able to get through this, because it can 'lower' its spines, but it will prevent interested dogs from using it. Perhaps cats might be inquisitive, but they are likely to beat a hasty retreat once they realise that they have a confrontation with an irate and very prickly animal!

Three 37 cm (15 in) pieces are nailed one above the other to 30 cm (12 in) pieces at each end to make a 30 cm (12 in) high back. The next step is to make the lid and the bottom, in which four pieces of 30 cm (12 in) long 10 cm x 18 mm wood (4 x $\frac{3}{4}$ in) are nailed to two pieces of 5 x 2.5 cm (2 x 1 in), ensuring that those for the floor go *outside* and those for the lid *inside*. The 10 cm x 18 mm (4 x $\frac{3}{4}$ in) overlaps 3.7 cm

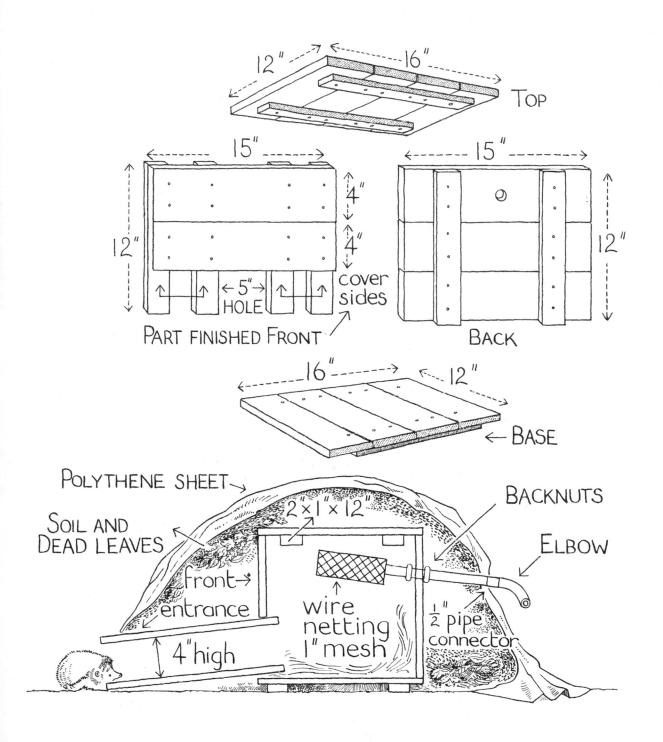

12" 16" TOP

15"
12" 4" 4"
5" HOLE cover sides
PART FINISHED FRONT

15"
12"
BACK

16" 12"
← BASE

POLYTHENE SHEET →
SOIL AND DEAD LEAVES
2" × 1" × 12"
BACKNUTS
ELBOW
front → entrance
wire netting 1" mesh
½" pipe connector
4" high

How to make a hedgehog house, as described on page 54 opposite

55

(1½ in) at each end, so that the lid fits. To make the front you will need 1.2 m (4 ft) lengths of 5 x 2.5 cm (2 x 1 in) – one at each end and two in the middle. These are fixed 12.5 cm (5 in) apart. Two 37 cm (15 in) lengths are nailed above the entrance. You should have one 30 cm (12 in) piece of 10 cm x 18 mm (4 x ¾ in) left, and from this saw off two 12.7 cm (5 in) lengths, and nail one to each side of the opening for the entrance tunnel.

Now drill a 12 mm (½ in) hole with a brace and bit in the middle of the top board at the back, before nailing the 0.9 m (3 ft) lengths which will make the side, and which have to be nailed to the front and back: the bottom has to be nailed to all four or vice versa! This hole is necessary to take a ventilator, which can be made of a 2.5 cm (1 in) pipe connector tapped to take two backnuts, and with a screwed elbow to fit on the outside. You should be able to obtain these pieces from a local ironmonger or plumber. It is useful to have a piece which is around 38 cm (15 in), so that there will be enough both inside the hedgehog's home and outside it. The end which is inside the house should be covered with 2.5 cm (1 in) mesh wire netting or perforated zinc. It is better if the hole is drilled in such a way that the pipe slopes downwards, so that it will drain any moisture from the hedgehog's house. The cover on the end of the pipe is necessary, because the hedgehog will bring in a great deal of debris, mainly leaves, which will fill the box. This would naturally block up the open hole.

All you need to do now is to place your box in a quiet part of your garden, but if you intend to do some hedgehog-watching, you will need to position the home in such a way that you can view it easily. The house needs to be camouflaged, and if you cover it with a sheet of polythene, ensuring that the ventilator is not obstructed, soil can be placed on top of this. You cannot be certain that hedgehogs will seek out what you might consider to be a 'desirable residence'. Nevertheless they might, especially if you place a piece of bacon rind inside your house to tempt them. You could also leave some hay outside so that if a hedgehog decides to move in, it will use this for bedding. The hedgehog will only take bedding if it is dry, so any which you leave, whilst being accessible, must also be covered.

This is likely to be a winter quarters for some lucky hedgehog. In the following spring it needs to be cleaned out, and to some extent fumigated, dur-ing the summer, ready for the following winter. Put a piece of bacon rind into the entrance to see whether it disappears: if it doesn't this indicates that the hedgehog is no longer in residence, and you can clean the house out. Pyrethrum powder or one of the herbal dog dusting powders should be used to try and ensure that any 'livestock' are removed.

Grey And Red Squirrels

One animal which is common even in town gardens is the grey squirrel. It's strange that this now-common British species was first brought to our shores as recently as the late 19th or early 20th century. Although you might like to see the animal in your garden you may regret it if you have trees, because it will damage them by stripping the bark away.

While the grey squirrel has had a population explosion, reaching pest proportions in many areas, the red squirrel is sadly on the decline. You would, therefore, be doing the animal a great service if you could persuade one, or preferably a pair, to take up residence. The red squirrel is declining so dramatically in many areas that even to locate the mammal proves an insurmountable task. The main requirement is that a garden should have enough mature trees to provide it with a good food supply throughout the year. It has been estimated that each red squirrel needs an area of two hectares (about five acres) which must have well-established pine trees to provide it with its food. Surely there are few gardens which can therefore support the red squirrel. But if you are in an area where the red squirrel still lives, you may find that it visits your garden, if you have the necessary pine trees. With an extremely fast turn of speed it will take to the trees, and be almost out of sight in a matter of seconds should impending danger threaten.

Bats And Bat Boxes

Although we can provide surroundings which are attractive to all sorts of wildlife, we can have no guarantee that anything will actually come and live there. But we can put out fruit for badgers and saucers of milk for hedgehogs, and likewise we can even put up nest-boxes for bats. Not unlike bird boxes in general principle, these roosting boxes will perhaps encourage the bats to stay in your garden. If you happen to have these visitors coming to your garden you will be aware that their excursions are seasonal. The bat hibernates during the colder months of the year and, nocturnal by nature when active, spends the day-light hours sleeping in suitable places, including buildings such as churches and in the hollow trunks of some trees. Increasingly hard-pressed as the number of trees decreases at an alarming rate, the bat will seek out any area which will afford it protection and in its search for suitable hibernatory quarters, this mammal will even make its way into caves and mines.

Unusual facts about bats' bodies
Mammals are 'warm blooded', as distinct from reptiles and amphibians, which are 'cold blooded'. Man, along with other mammals, has a body temperature which stays fairly constant. Even when he is ill it only rises by a few degrees. The bat is extraordinarily different. Its body temperature fluctuates widely, depending on the activity in which it is engaged. Like cold blooded animals the bat tends to

Grey squirrel on turkey oak

57

British bats: (a) the common bat, (b) the great bat, (c) the long-eared bat

take on the temperature of its surroundings, particularly when at rest. Indeed it is not uncommon for it to be near freezing point (0°C/32°F) during its period of hibernation in winter. This is a particularly valuable adaptation, since it means that the animal will be using up only very small amounts of energy, thus conserving the supply of fat which it has built up before it hibernated. In actual fact the bat's temperature is usually around 40°C (104°F). Perhaps what is more remarkable is its heart beat. In flight it may go up to *1000 per minute*; man's is usually around 150 during the height of activity, and about 70 when performing routine activites. The interesting fact is that within a short time of landing the bat's temperature will have dropped to about 10°C (50°F), and within half an hour of expelling its last droppings, the body temperature will normally be the same as that of the bat's immediate environment. A typical temperature in Britain is between 16° and 21°C (60.8° and 69.8°F).

A place to hibernate

Towards the end of October, and in some colder years perhaps a little earlier, the supply of insects, the bat's chief source of food, decreases rapidly, and so the mammal will go into hibernation. During the preceeding months, like the hedgehog, the bat has been eating more than is necessary, and the excess is stored as fat, which will serve two purposes during the winter: it will act as an insulating layer and so protect the bat, to some extent, from the cold; in addition the fat acts as a store of food, which is gradually used up throughout the winter months. Although a hibernating bat may be taken for dead its vital body activities, whilst at a very low level, continue. In the erratic climatic conditions of the British Isles it is not unusual for the temperature to rise quite sharply to give mild, almost spring-like conditions during winter. If these temperatures persist for a while the hibernating bats tend to stir and are tempted to take to the wing, their inner mechanism misinforming them that spring has arrived. Lured from this rest they will not find the insects which they seek, and the very activity will quickly reduce their supply of stored food. Should the mild conditions continue and the bats still search for food, the reserves of fat could become very low. If this happens it is likely that the animal will die during the hibernatory state to which it will return. If disturbed during hibernation it is not uncommon for a bat to take up to an hour to 'warm up' as it were, and be ready to leave its winter's quarters.

Bat roosting-boxes

Boxes for birds to build their nests in have been used for quite a while, and a good deal of information has been collected which enables expert organizations to assess the best size of boxes, type and size of hole *etc*. Strangely enough, although bats often live in close proximity to man, such detailed information is not available when it comes to providing bats with roosting-boxes. Conservation is now necessary however, because the bat population has declined to such an extent that two species, the Greater Horseshoe Bat and the Mouse-eared Bat, have had to be legally protected under the Conservation of Wild Creatures and Wild Plants Act 1975.

Because so many people regard bats with a mixture of superstitious fear and simple disdain, the Society for the Promotion of Nature Conservation (SPNC) has produced an excellent illustrated leaflet on their conservation and control which sets down guidelines for putting up bat roosting-boxes in your garden. As bats prefer a rough surface on which to alight and crawl about, bat-boxes should be constructed from thick rough-sawn timber which should *not* be treated with preservatives, as strong smells repel them. As shown in the diagram, a slit of about 15 mm should be provided at the bottom of the box to allow them to get in whilst minimizing draughts. The box should be about 10 cm square inside and indeed, up to fifty bats have been discovered in a box measuring 10 x 10 x 10 cm (4 x 4 x 4 in)! As so little is known about the precise artificial roosting requirements of bats, a number of boxes should be made and fixed to trees at different heights and aligned in different directions. Heights should vary from 5 feet upwards, and it has been noted by the SPNC that boxes facing south-east, allowing the sun to fall on them, are preferred in summer, while boxes with a northerly aspect may be used for hibernation.

Patience is inevitably the keynote – as with watching all wildlife – because it has been estimated that it may be as long as *three years* before a bat will take up residence in a confined box. Unfortunately if you wish to attract bats to your garden there is virtually nothing else you can do except put up your boxes

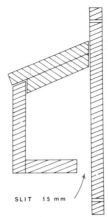

SLIT 15 mm

A simple bat-box to make

and wait, because unlike other mammals who may be attracted by the provision of special food, bats feed on large numbers of insects caught during flight and the only real necessity with which man can provide them is a safe place to roost. Originally a creature of the woodland where it was able to establish itself in naturally hollowed-out trees, with the loss of vast areas of wooded sites, the bat has had to change its habits. Now that bats will seek sanctuary in man-made places, your nesting boxes may prove to be life-savers and indispensible to them, as they continue their search for new places to roost.

The Mole

Sleek, and fitted for its subterranean life, the mole is a fascinating creature. Although not appreciated by most gardeners, it is valuable because in making its underground burrows it is constantly mixing up the soil. Apart from a period during its early life, much of its time is spent below the surface of the soil, where it tunnels relentlessly and does a very useful job turning up the lowest layers to be mixed with the decaying material in the upper layers.

The mole's activities are clearly to be seen in the form of the regular mounds of earth scattered over fields and in gardens, simply known as mole hills. The soil of which these heaps are composed is usually a different colour, because the mole has brought it up from a lower layer. In addition to its deep tunnelling activities, the mole also works a few centimetres below the surface, and instead of digging away and shovelling the soil backwards, it just pushes it up in a ridge of soil. In this way its

The mole

movements can be followed quite easily. The mole has a daily routine and it sticks to this with almost religious fervour. During each eight hour period the mole works for four and a half hours and rests for three and a half hours. Although it spends the greatest part of its life under the ground it will come to the surface in search of food on damp nights. This is about the only time that you will actually see the creature, whose presence you have been aware of perhaps for a very long time. If you wish to attract the mole to your garden, the best method is simply to ensure that your soil is healthy and full of earthworms, as these, together with grubs are the mainstay of the mole's diet.

The Badger

Unless you are lucky enough to have a very large garden, you are unlikely to have Brock the badger as a resident. However, even in areas of dense population, the badger may be an occasional, even a frequent, visitor.

The by-word for the badger's life-style is 'caution'. It is perhaps rather strange that such a large animal, with no known enemies – except of course every animal's enemy: man – should be so wary of leaving its sett and venturing forth. Nevertheless, the badger is distinctly a creature of habit, and once it has made a route from its sett to another area it will use this track continuously, until it has a well-worn pathway, along which it will probably stroll most evenings when the weather is suitable. If we exclude the deer, the badger is one of the largest of our so-called native animals. We are told that it is originally a native of Africa, but it is thought to have become 'European' as early as the Stone

Age, when our island was part of a much larger continent.

There is little you can do to encourage the badger to visit your garden, and even if he is already there, there's not much you can do to ensure that he stays. Undoubtedly if he is resident, you will be aware of his presence: if he is a visitor the signs might not be so obvious, but look out for the badger's tracks, which are distinctive as you will see from the illustration on page 63.

As a stickler for hygiene the badger probably has few equals in the animal world. Latrines are dug away from the sett and these are used regularly so that the badgers' home is not soiled. Dirty bedding is removed as necessary, and new material is collected and taken to the sett to replace it. Having amassed a supply of suitable material Brock shuffles backwards towards his home, the bedding carried in his front paws. The badger always selects a tree trunk near the sett for use as a scratching post, and after depositing the clean bedding in its home, he cleans his claws on the post.

Encouraging and observing the badger

Obviously you are only going to have badgers visiting your garden if they happen to be in the neighbourhood. The animal's distribution is by no means uniform over the British Isles. Woodland is a favourite haunt, with banks and ditches also being sought out. Once established the badger's sett tends to grow. This makes the badger unpopular in some areas and gassing is sometimes carried out to control its numbers. This is quite illegal except when the animal is known to spread bovine tuberculosis and has to be put down to prevent the disease from spreading.

Having established the fact that there is a sett near your home, you will need to keep watch. If the mammal is a regular visitor you will, as we have already mentioned, probably find a well-worn track into your back garden. If this is so you can leave some fruit here, and if you can produce suitable hides in the area you will be able to observe the animal. You will need to ensure that the lookout is downwind, so that the badger does not get scent of you, otherwise he will be off. Your vigils might not always be rewarded, because any kind of disturbance as he is about to leave the sett will deter him, and in an effort to avoid any would-be danger Brock will stay indoors, rather than venture out.

Brock the badger

How To Make A Hide

You might consider that going to the lengths of providing a hide to observe birds and animals in your garden is not really justified. However, movements which you make when trying to watch your garden visitors are likely to perturb and frighten many birds and animals and a hide is something which will soon become an accepted piece of your garden furniture. Even if it is portable, and you plan to move it about, the animals will still get used to it. From its security you are likely to be able to discover a great deal about the wildlife in your garden.

Common shrew

You can buy a hide from a commercial supplier (see page 65) or make your own. A framework can be made from either dowel or broom handles. If you sharpen one end of each of four pieces of wood they will go into the ground more easily. Strong wire is used to form the top of the frame and make a box-like structure. You can use small spring clips or screw-type eyes for attaching the wire to the uprights. The 'hook' that fits into the eye is made by bending over the end of the wire. You will need some material to cover the hide. It is likely that you will want to use it in wet weather, in which case you will need waterproof material. Polythene can be used, but it has the disadvantage that it is noisy in the wind. If you make the material into a 'box', with the bottom left open, this will fit easily over your frame. Alternatively you can sew up a couple of the sides, and use velchro or a zip to complete the rest. Peepholes need to be cut in all four sides, and flaps sewn over the holes, so that they will not disturb the birds or other animals when they are not being used. Of course the advantage of a portable hide is that you can easily take it into the countryside as well.

SOME OF BRITAIN'S MAMMALS LIKELY TO OCCUR IN GARDENS FROM TIME TO TIME

SPECIES	FOOD	BREEDING	DISTRIBUTION	IDENTIFICATION
PYGMY SHREW (*Sorex minutus*)	Seeks out beetles, insects and spiders, *etc.*, especially in leaf litter. Will also take worms and sometimes carrion	Mates from about April to July. Between 2-8 offspring. Does not live for many months	Found over much of Britain. Our smallest mammal	About 5-6cm (2-2.5in) + 3.5cm (1.5in) of tail. Has brownish hair with grey above. The underside is greyish-white
COMMON SHREW (*Sorex araneus*)	As for Pygmy shrew, but usually takes larger insects, *etc.*	Mates in spring through to summer. Usually two litters of 5-10 youngsters. Thought to live less than a year	Found over much of the country, except Ireland	About 7-8cm (3in) long + 3.5-4cm (1½in) tail. Small protruding eyes; pointed, whiskered snout. Rusty-brown coat, varies pale to darker. Yellow-grey on underside
HEDGEHOG (*Erinaceus europaeus*)	Insects, worms, slugs, snails, frogs, small animals, *etc*.	Mates between March and July. Young (3-8) born after 31 days. Have pink spines: eyes open after 14 days	Widespread over Britain – less common in the Highlands	A plump body covered with spines. Face area has coarse hairs
COMMON MOLE (*Talpa europaea*)	Mainly soil living animals, including earthworms and insects	From March to May: young are born after 3-4 weeks. There are from 2-6 blind and naked young. Eyes open in 21 days	Occurs widely over most of the country	Has a barrel-shaped body: the fur is close and compact, and gives the appearance of velvet
PIPISTRELLE BAT (*Pipistrellus pipistrellus*)	Mainly small flying insects	Little is known about its breeding habits	It is the most common and the smallest of the British bats, and very widespread	About 7-8cm (3in), including tail. Wing-span of 18-20cm (8in). Colour varies from yellow through to rust-brown, and even dark brown: sometimes black
SEROTINE BAT (*Eptesicus serotinus*)	Mainly moths and beetles	Little is known about its breeding habits	Occurs mainly in Southern England	12.5cm (5in): wing-span of 35 cm (14ins). Brownish above brownish-yellow below
NOCTULE BAT (*Nyctalus noctula*)	Mainly night-flying moths and other nocturnal insects	Little is known about its breeding habits	Quite common in England and Wales: not found as often in Scotland	12-13cm (5in); narrow wings about 35-40cm (14-16in). The fur has a silky texture and is generally reddish-brown on the upper surface; a little paler underneath
BADGER (*Meles meles*)	Takes a wide range of plant and animal matter. Is fond of invertebrates, including worms, as well as small mammals, plants bulbs, corms, *etc.*	Mating takes place early in spring and may continue into summer. Young – 3-5 – born the following February-April. Lives for about 15 years	Widespread, although often unknown, because of its nocturnal nature. Prefers sand and chalk soil, especially in wooded areas cially in wooded areas	Large body with length of 80cm (32in). The fur is a mixture of grey and whitish hairs, intermingled with black ones. The sides of the body are lighter. The 'distinguishing' feature is the black and white markings on the face the face
WEASEL (*Mustela nivalis*)	Mainly mice and voles, also other small mammals: takes birds and some eggs	Pair and mate between April and August. Young born 6 weeks later, with between 5-10 kittens in a litter. Usually 2 litters a year. Lives for 7-10 years	Found over much of Britain, although is becoming rarer in some areas	Female about 17cm (7in) + 5cm (2in) of tail: male 20cm (8in) + 7.5cm (3in) tail. The coat varies from red to brown above and it is white on the undersurface
RED FOX (*Vulpes vulpes*)	Wide range, mainly birds/small mammals; also some insects	Mates in January or February. After about 60 days a litter of 3-6 cubs are born. The red fox lives for about 12 years	Still common in most parts of the country	About 60cm (24in) + 40cm (16in) of tail. Generally chestnut brown. The underside and chest are ash-grey. Tip of the tail, underside of head, inside of legs and neck are white
RED SQUIRREL (*Sciurus vulgaris*)	Seeds, particularly conifers, as well as shoots fungi, acorns, nuts, berries, *etc.*	Mating takes place from spring to summer. Between 3½-5 weeks later 3-7 young are born. There may be up to 5 litters in any year	Disappearing – found in East Anglia, Lake District, Wales, Devon and the Highlands	20cm (8in) in length + 19cm (7in) tail. Generally reddish-brown above, with white on the underside
GREY SQUIRREL (*Sciurus caroliniensis*)	Nuts, seeds, fruits, shoots, bulbs, bark, roots, sometimes insects, and birds eggs. Will bury food	Mates from early spring to middle of summer, although the greatest mating appears to take place December-January and May-June. About 44 days later, litter of 1-6 born	Found in most parts of Britain	25cm (10in) + 20cm (8in) tail. In summer generally brownish-grey above with coloured hairs on flanks. Underside white. In winter the coat thickens and becomes greyer; tail has more white hairs to fringe it
BLACK (HOUSE OR SHIP) RAT (*Rattus rattus*)	Will take almost anything it can find	May mate anytime from spring through to autumn. Young of between 8-20 will be born 3 weeks later. Lives for 6-7 years	Now quite rare in some areas. There are 3 subspecies	Length: 17cm + 19cm (7.5-8in) of tail. May be almost totally black, but varies through to almost light brown
BROWN (COMMON) RAT (*Rattus norvegicus*)	Will take what is going, including small animals, as well as cereals	Mating takes place from March to September/October. Young born after 24 days: may be 2-7 litters a year, with between 5-20 in each. The young are naked. Lives for 3-5 years	Occurs over most of the British Isles, but usually absent on high mountains	Is much larger than the black rat, with a much thicker body, 20-22cm (8-9in) in length, with tail rather less than body length. Coat varies from greyish brown through a reddish tinge: off-white underneath
HOUSE MOUSE (*Mus musculus*)	Will take almost anything which is available and edible	Mates for almost all Year – early in spring to late in autumn. Between 5-10 litters a year, after gestation of 19-20 days: 4-8 young in a litter. Lives for between 2-4 years	Occurs almost anywhere where food is found, as long as there is some shelter as well	Length: 7.5cm (3in) + 10cm (4in) tail. Generally grey – although the underside is lighter than the upper
HARVEST MOUSE (*Micromys minutus*)	Insects, berries, fruits and seeds	Mates any time from April to September. A litter of between 3-9 naked young born 21 days later: 2-3 litters a year	Now rare in some areas, partly due to modern farming methods	Length: 5.5cm (2¼in) + 5cm (2in) of tail. Above the coat has a brownish tinge; below it varies from a yellowish to greyish
WOOD MOUSE (Long-tailed Field Mouse) (*Apodemus sylvaticus*)	Fruits, berries, nuts, grain, buds, peas, etc. and some small insects	Mating occurs from early in spring through to autumn: 3½ weeks later 3-8 young are born. Lives for between 2-4 years	Widespread in many habitants	Length: 9cm (3.5in) + 7.5cm (3in) of tail Greyish-brown above, with longer reddish-brown hairs: grey-white on undersurface. Usually has a yellowish chest patch, which varies from animal to animal

SPECIES	FOOD	BREEDING	DISTRIBUTION	IDENTIFICATION
BANK VOLE (*Clethrionomys glareolus*)	Berries, seeds, buds bark and some insects	Mating takes place from April to September. After 18 days, between 3-5 young are born: 3-4 litters each year. Lives for 2-3 years	Fairly widely distributed over much of the country	Length: 11cm (4½in) + 5cm (2in) tail. Coat varies considerably from a rust-brown to darker brown. The underside is grey to white
COMMON VOLE (*Microtus arvalis*)	Bark, seeds and grass	From March to October: 4-8 young are born 3 weeks later. Lives 2-3 years	Widely found with various races having been identified	About the same size as field vole: tail slightly longer
FIELD VOLE (Short-tailed Field Mouse) (*Microtus agrestis*)	Mostly grass, although does feed on roots and bark from time to time	March to September/October: 4-7 in litter; 3-4 litters a year. Lives for 2-3 years	One of our more common mammals	Length: 10 cm (4in) + 2.5cm (1in) tail. Mixture of black and grey brown hairs Rather lighter, tending towards grey/off-white below
RABBIT (*Oryctolagus cuniculus*)	Grasses and other plants, including many cereal crops	Mating takes place from February to June: 4-12 young; 3-5 litters. Lives for 5-7 years	Although decimated by myxomatosis, it is now on the increase	Length: 40cm (16in) + 7.5cm (3in) tail. There is a mixture of brown, grey and yellow hairs. The underside is whiter, as are the chest and inside of the legs. Underside of the tail is white

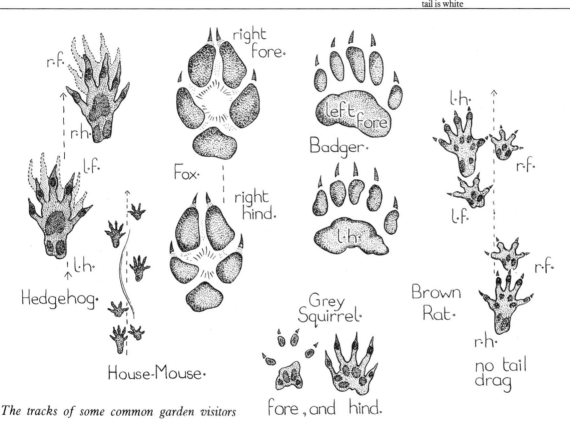

The tracks of some common garden visitors

Further Useful Information

Organizations and Societies

Mammal Society, Harvest House, 62 London Road, Reading, Berks, RG1 5AS

Hon. Secretary Prof. M Delaney, School of Environmental Science, University of Bradford, Bradford, W. Yorkshire, BD7 1DP

The Mammal Society is a national organization which brings together everyone, professional or amateur, interested in mammals. It publishes various journals, leaflets, *etc.* There is a quarterly newsletter. Various groups, like those concerned with the study of bats, form part of the Society.

Henry Doubleday Research Association, Convent Lane, Bocking, Braintree, Essex

The HDRA is *not* a wildlife organization, but is concerned with gardening and farming: their booklets will prove of interest to those people who are concerned about the effects of pesticides, *etc.* Membership costs £5 per annum. Further details available (please enclose s.a.e.).

Periodicals and Magazines
Mammal Review, Mammal Society, address earlier
in this section

Books, Leaflets, etc
Bang, P. and Dahlstrom, P., *Animal Tracks and
Signs*, Collins

Domestic mice

British shrews
Brink, F. H. Van den, *Mammals of Britain and
Europe*, Collins
Burton, M., *Observer's Book of Wild Animals*,
Warne
Burton M., *Wild Animals of the British Isles*, Warne
Burton, M., *The Hedgehog*, Andre Deutsch
Corbet, G.B. and Southern, H. N., *The Handbook
of British Mammals*, Blackwell

Devon Trust for Nature Conservation, *In Defence of
Bats*
Duerden, N., *Mammals of Great Britain*, Jarrold
Friends of the Earth, *The Declining Otter*, FOE
Otter Campaign, Yew Tree Cottage, Caffcombe,
Chard, Somerset

The rat

Hanney, P., *Rodents*, David & Charles
Harrison Matthews, L., *British Mammals*, Collins
Lawrence, M.J. and Brown, R.W., *Mammals of
Britain, their Tracks, Trails and Signs*, Blandford
Lockley, R.M., *Private Life of the Rabbit*, Andre
Deutsch
Lyneborg, Leif, *Mammals in Colour*, Blandford
Mellanby, K., *The Mole*, Collins

Young foxes

British moles in their native haunts

Some British weasels

Morris, P., *Hedgehogs*, Forestry Record 77, Forestry Commission/HMSO

Neal, E. G., *Badgers in Woodlands*, Forestry Commission/HMSO

Neal, E., *The Badger*, Collins (also in paperback)

Nixon, M., *Oxford Book of Vertebrates*, Oxford University Press

Operation Tiggywinkle, Henry Doubleday Research Association, *Information about the Hedgehog*

Head of the wren

Rogers, Brambell, *Voles and Field Mice*, Forestry Record 90, HMSO

Shorten, M., *Squirrels*, Collins

Society for the Promotion of Nature Conservation, *Focus on Bats — A Guide to Their Conservation and Control*, SPNC, 2 The Green, Nettleham, Lincoln

Southern, H. N., *Handbook of British Mammals*, Blackwell

Vesey-Fitzgerald, B., *Town Fox, Country Fox*, Andre Deutsch

Suppliers

Ready-made 'hides' from which you can observe the wildlife visitors to your garden can be obtained from: Jamie Wood Ltd, Cross Street, Polegate, Sussex

BUTTERFLIES, MOTHS, BEES AND WASPS

The illustration shows (anti-clockwise, from top):
buddleia with peacock resting, red admiral at
the brambles, small tortoiseshell, comma butterfly
with wings closed on blackberries, honesty
and nettles growing by old potting shed

Butterflies And Moths

The term 'bird gardening' was coined some years ago and 'butterfly gardening' is also a viable idea because with our efforts we may be able to encourage these magnificent insects to visit, breed and over-winter. What we are able to do will depend on the size of our estate! But of course the greater the variety of plants which we are able to propagate and grow, the greater the probability of encouraging a wider range of butterflies.

Naturally it is necessary to isolate the requirements which will attract 'resident' butterflies. In all honesty it must be said at the onset that although it is not difficult to encourage butterflies to breed, however carefully one plans the butterfly assignment, these insects will only come to the garden if they happen to be in the vicinity anyway, and there is no guarantee that they will come to stay even if you provide a suitable habitat for them. Initially you will need to know which species frequent your locality so that you can plan the food plants which you need to provide. A garden planned for butterflies is more likely to attract them than a garden where no special provisions have been made.

The butterfly life-cycle

If we understand something about the life of butterflies, it becomes easier for us to plan for their needs. Since all butterflies have the same stages in their life-cycle, the following is applicable to all species. The times when you will find eggs, caterpillars, pupae and adults (or imagos) will vary with the species, and the table on page 71 summarizes the timetable for most common butterflies. In the life-cycle of each and every butterfly there are four stages. The female will lay her eggs after mating has taken place. Since it is the warmth of the sun's rays which will encourage the eggs to hatch, these have to be laid in sunny weather. If during the 'incubation' period, from the time when the eggs are laid to the emergence of the caterpillars, the weather turns damp and cold the eggs may take longer to hatch, and a prolonged period of unfavourable weather may cause them to perish.

The life span of an adult may be little more than two weeks. This is the average: some live longer, others for shorter periods, although some of those which hibernate have a much longer lifespan. So after mating the eggs must be laid as quickly as possible.

The eggs of all butterflies are very small, most seldom measuring one millimetre in either height or width. Although there is a wide variety of pattern and form to the eggs of the various species, the butterflies belonging to the same family have similar markings and shape.

The common butterflies

Using scent as a guide the female assiduously seeks out a plant which will provide the vital food for her offspring, the caterpillars, when they hatch. When she thinks that she has found a suitable plant she lands on it, and then, using her front feet, she continually 'kicks' the leaf. In this way she is able to decide whether the leaf is the right one for her to lay her eggs on. Once she is certain that she has found the right plant she will lay her eggs.

It is important that if the eggs are not actually laid on the food plant, they should be laid close by. Some females lay their eggs singly, placing one on each plant. Some, like the small tortoiseshell and the peacock, lay theirs in batches. Once the eggs are laid, the length of time which passes before the caterpillars emerge varies considerably. Sometimes the process is complete within six days and in other cases it may be eight months.

The egg hatches and the caterpillar, or larva, emerges. In many species its first task is to eat at least a part of the eggshell from which it has just emerged. Then it will feed on the plant which provides its supply of food until it is fully grown. There are some species which hatch, eat the eggshell and then go into hibernation until the following spring.

Because the growth of the larva is limited by the size of the skin, this is discarded at regular intervals. Underneath will be a new, larger skin. During moulting – in some species it may happen three times during the caterpillar stage and in others up to six times – the caterpillar is immobilized and is naturally the target for various hungry predators. At this time, it is possible to see why it is necessary to have such a large number of eggs in the first place!

A typical insect, the caterpillar has three pairs of legs. At first glance this seems confusing, as there appear to be six pairs – one on each of the first six

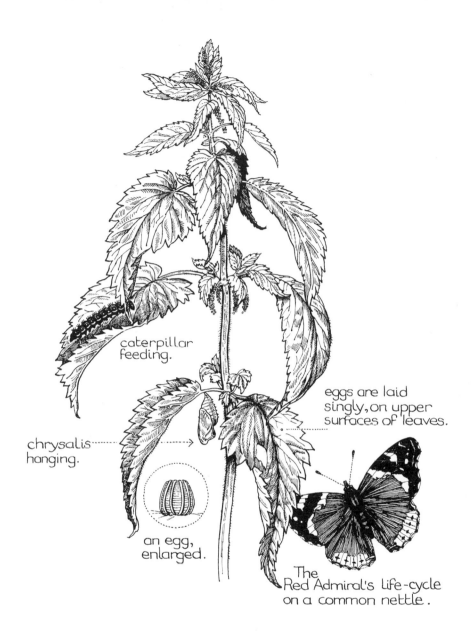

caterpillar
feeding.

chrysalis
hanging.

an egg,
enlarged.

eggs are laid
singly, on upper
surfaces of leaves.

The
Red Admiral's life-cycle
on a common nettle.

The butterfly life-cycle

segments. However, it is only the first three pairs which are true legs and the others are known as prolegs or claspers. There is also a pair of claspers on the final, anal segment. These enable the caterpillar to resist its removal from its food plant by a hungry predator. The larvae of the sixty-nine British species vary considerably in shape, colouring and size. Having grown to its full size, the caterpillar no longer needs any nourishment, because it is now ready to pupate.

The pupa or chrysalis represents the third stage in the butterfly's life-cycle. Although a remarkable change will take place within the chrysalis, this stage of its life is one of total immobilization, apart, perhaps, from a few twitches or wriggles, should the case be disturbed. As the time for emergence draws

Small copper

Large skipper at rest

near it is possible to see the outline of the developing butterfly within the pupal case. In some species it is even possible to detect the colour of the butterfly inside its protective shell.

Once the development has taken place inside the pupal case, the adult or imago, is ready to emerge. The case will split and the adult will be free. At first sight the insect bears little resemblance to our image of, for example, a red admiral or a swallowtail. The wings, besides being limp and floppy, are also much smaller than in the adult ready to fly. For perhaps the next hour or so the butterfly will remain where it has emerged, the veins of the wings slowly filling out and ultimately becoming 'set'. Once the wings are filled completely with liquid, the newly emerged adult butterfly will put them to the test. Life is very short, and time is at a premium and the next important task is immediately to continue the species by mating.

Speckled wood

Over-wintering

Depending on the particular species concerned, butterflies spend the winter in one of the four life-history stages. Some spend this period as eggs, some as caterpillars, others as pupae, and there are five species which hibernate as adults. The latter are the large tortoiseshell – a rare species, the small tortoiseshell, a common garden visitor, the peacock – also seen around our houses, the brimstone and the comma. In addition, the red admiral *may* manage to survive the winter, but the numbers living through to the following spring will be small. Having taken their fill of nectar in the autumn from flowers in the garden, these butterflies will search out their winter quarters. If you have a hollow tree in your garden, you may find peacocks settling there to hibernate.

On the other hand the small tortoiseshell, masses of which have probably covered your autumn-flowering michaelmas daisies, will go for buildings. Outhouses such as sheds and garages, as well as barns, are favourite places. Many, too, manage to find their way into houses. The older houses with sash windows seem popular. I have collected large numbers of these insects from an old house in which I lived as a boy, and released them into the garden as they woke from their winter's rest. Many had sought out the crevices around the old-fashioned windows. Strange as it may seem, only recently a number flew out from a divan bed, where the hessian had come away from the underside!

Should an unduly warm spell in winter penetrate into the butterflies' hibernating quarters, they may be stirred to come to sun themselves, or perhaps even to search out a suitable supply of food, although the chances of finding any are not very high. They do not generally seem to wander far from their sleeping sites, and will often be seen, their wings outstretched, clinging to the wall of an outbuilding.

Large heath

SOME COMMON BRITISH BUTTERFLIES AND THEIR HABITS

SPECIES	FOOD	PERIOD OF ADULTHOOD	EGGS/LARVAE/ PUPAE
Brimstone	Buckthorn	All, except June	May-August
Gatekeeper	Almost all grasses	July & August	All, except July/August
Large white	Cabbage, nasturtium	April-August	All, except July/August
Peacock	Stinging nettle	June & Aug-Dec	June-August
Small blue	Kidney vetch	May-June & Aug-Sept	Jan-Dec
Small heath	Almost all grasses	May-August	Jan-Dec
Small white	Cabbage/Hedge mustard	April-August	Jan-Dec
Small tortoiseshell	Stinging nettle	All year	May-Sept
Wall brown	Prefers annual meadow grass: will take others	May-August	All year
Orange tip	Honesty/Lady's smock	April-June	All year
Ringlet	Cockfoot and other grasses	July	All year
Small copper	Sheep's sorrel	April, June-Sept	Jan-Mar, May-June August-Dec
Green-veined white	Garlic mustard	April-August	All year
Comma	Wych elm, nettle hop, currant, gooseberry	All, except June	May-Sept
Common blue	Birds-foot trefoil, black medick	May-Sept	Jan-April, June-July October-December
Speckled wood	Grasses	April-May July-August	Jan-April, July-Dec
Small skipper	Various grasses	July	All year

Elephant hawk-moth

The moth life-cycle

The general difference between butterflies and moths is that the former are diurnal or day-flying and the latter nocturnal or night-flying, or so it is claimed. Another characteristic of moths is that they settle with their wings in the folded position, and that they have 'thick' bodies, which are often more hairy than butterflies. Whilst such characteristics hold good for the majority of species, there are as usual bound to be exceptions to the rule and not all moths fly by night: there are a number which are on the wing during the daytime. There are also exceptions to the rule about body shape because some butterflies have thick hairy bodies, including the various species of skippers, and fold their wings while at rest! On the other hand, there are many species of moths which have slender butterfly-like bodies. However, butterflies found in the British Isles do have one distinguishing feature. At the end of the butterfly's 'feelers' or *antennae* there are small knobs (the antennae are 'clubbed') and this characteristic is missing from moths.

Although we know a great deal about the seventy or so species of butterflies in the British Isles, relatively little is known about the two thousand or more species of moths. Yet large numbers of moths are likely to take advantage of your garden territory for their egg-laying activities, and many different species of caterpillars should be encountered.

Of the more spectacular moths are various hawk-moth species. Perhaps one of the most interesting is the elephant hawk-moth whose life-cycle is given here. Although a large species, it does not get the name 'elephant' from its size, but from the fact that in the caterpillar stage the body tapers, and

when the head is extended one gets the impression of an elephant's trunk. The larvae are instantly recognizable in their later stages. About 7.5 cm (3 in) in length, they have a dark ground colour, which varies between brown and black. This is highlighted with pink and white spots, which are found on the second and third segments. It is this which some people consider gives the insect a grotesque appearance, but for the moth its appearance is its only defence. If this fails to frighten off enemies the caterpillar is likely to succumb to predators, because it has no other means of defending itself. In its early days the

Burnished brass

larva is green, and it is not unusual for it to remain this colour even in the fully grown caterpillar.

The larvae are generally encountered in the summer months, from around July through to September. Most females will lay their eggs on either rosebay willow-herb or the great willow-herb. When this is not available other plants are sought out, including balsam, fuschia, enchanters' nightshade and marsh bedstraw. As the time arrives for the caterpillars to pupate they move down the plant stem to the soil, where they will usually make a 'home' by combining soil and leaves together, or simply bury themselves in the soil. Here they will stay until they emerge as beautiful insects. The adults can usually be seen on the wing from about June. If many moths are missed because of their insignificant nature, this cannot be said of the

Lime hawk-moth

elephant hawk-moth; they have a wingspan of between 6 and 7.5 cm ($2\frac{1}{2}$ to 3 in), and the pink, so obvious in many of the caterpillars, is also apparent on the adults. In some it is pink: in others it tends towards a purplish tinge.

These are not the only moths which are spectacular. Watch out for the larvae and adults of other hawk-moths, particularly the privet hawk-moth, whose life-cycle will also form a fascinating subject for the wildlife gardener.

Red underwing

How To Breed Butterflies

The chances of success when attempting to breed butterflies are not always as great as one would hope. Because the caterpillars are the most active of the pre-adult stage, they are often the easiest to find to begin your breeding. In order to find them you will need to know the food plant of each common species. For example, if you want to collect the larvae of the small tortoiseshell, it is little use looking on dock leaves. If you refer to the small tortoiseshell life history given on page 81 you will also discover that it is necessary for the stinging nettles on which they breed to grow in particular situations. Because of the communal nature of this and some other species, you are more likely to locate them than those which live a solitary life. Treat the caterpillars with great care. Very small ones can be picked up with a paint brush.

Wall brown

Looking after caterpillars
Once you have found your insects you will need to know something of their requirements, and especially their food. If you are looking for specific caterpillars you will probably have some idea of their needs. If you come across a caterpillar you do not know, you will need to consult a reference book, (see page 94).

Cages for the caterpillars can be bought from a supplier, but are generally expensive. You can make your own, using wood or cardboard. You will need a container which has a removable lid, and preferably with at least one side – the front – replaced by any net with a small mesh. If you are using a wooden or bought container that is to be re-used you will need to ensure that once the caterpillars have left, the interior is disinfected with a dilute solution of sodium hydroxide. Caterpillars are likely to suffer considerable deaths from disease. Cardboard cages are, therefore, preferable, since they are expendable, and the chance of transmitting diseases is much lower.

Silver washed fritillary

Providing food

If you haven't got the necessary plants growing in pots already you can either dig up the plants (but see the Law for Wild Plants, page 6), or use cut stems, remembering that in order to feed easily caterpillars need fresh leaves. Since there are often large numbers of small invertebrate animals on the leaves of both potted and cut plants, disease is likely to be a problem. To rid the plants of their 'guests' they should be laid in a bowl of water for ten minutes or so. Virtually all the inhabitants will escape from the leaf in an effort to avoid drowning. If using a pot plant it is a good idea to cover the soil with polythene, since there are many pests in the soil which might come out and infect the caterpillars. If cut stems are being used it is necessary to make sure

that the top of the container in which they are placed is sealed with cotton wool, to prevent the caterpillars from drowning.

Potted plants should be replaced as soon as they show signs of depletion: cut stems should be taken out each day, and newly collected ones put in. Removing the caterpillars from the old plant might prove rather a problem. One way of overcoming this is to have a cage which is big enough to keep both pots in for a short while. The larvae will usually move quite quickly from the old to the new plant.

Mention has already been made of the serious threat of disease. There are other ways in which it can be minimized. First, it should be a rule to handle the caterpillar as little as possible. Second, overcrowding often brings about a greater disease risk. Several small cardboard cages are better than one large one.

Common blue

Comma

Peacock

Honesty

Painted lady

From larva to pupa

There are vast differences in the action which various caterpillars take before they are ready to change, as it were from juvenile to adult. In the natural state many go to the ground; others make 'tents' (see page 70), and so on. In a cage they all usually move towards the top of the container. They will successfully make the transition from caterpillar to pupa to adult here. Opinions as to whether you should leave them or remove them vary from one entomologist to another. Personally, I feel there is no reason why they shouldn't be left, as long as the cage is secure, and they are not likely to be preyed upon. If you feel happier about moving them they should be taken out when they have moved upwards. A piece of gauze on the inside of the lid of a cardboard box will ensure that they have a surface to which to cling.

Emergence of the adults

I always feel that it is wiser to leave the caterpillars in the cage, because there are always difficulties as to when they are likely to change into adults. However, one or two tips are given here. Up to a month before they are likely to change they should be placed in an indoor cage with twigs, etc. Some people recommend that the pupae should be gently sprayed with some warm water. This is a question of anticipating emergence, because this operation should only be undertaken a few days before the butterfly is due to come out. Making a decision about this may be easy for the trained entomologist, but for the novice and layman it is a different matter.

Once the pupal case has split, and the adult has emerged, it will take a time for the wings to become filled with liquid. This accomplished, some butterflies might want a supply of food. It is a good idea to have a cage in the garden, which covers a nectar-rich flower. Care should be taken to ensure that the butterflies are not likely to suffer from direct exposure, either to excessive heat or to rain.

Sleeving caterpillars

An alternative way of retaining caterpillars which are feeding in your garden is to leave them on the plant, but to protect them by sleeving. This is not a difficult operation.

You will need to cut a piece of sheeting which is between 60 and 90 cms (2-3 ft) square. A tube should be made from this by sewing along the length of the longest side. You now need to find caterpillars on a plant. Slide the sleeve over the specimen, and tie both ends. You should check that the ends are secure, and make a regular inspection. It, and the caterpillar should be removed once the food supply has been exhausted. The greatest disadvantage is that you will not be able to observe your caterpillars, as you would if they were in a cage.

Releasing your butterflies

Once you have attracted and bred butterflies in your garden, you will perhaps think it a good idea to release them into the wild, to allow them the freedom of their natural state. But don't do it until you have carefully considered the idea, and at least consulted your local trust for nature conservation to make sure there is nothing to advise against this in your area. Most butterflies are specific about the food plants which they need, and to release these insects into an area where the food plant is either absent or scarce will almost certainly court disaster, and result in the loss of the species which you have so carefully nurtured. The British Butterfly Conservation Society has had a certain amount of success with increasing stocks of some species. They will be only too pleased to offer suggestions and advice about your butterfly conservation efforts, but please enclose a stamped addressed envelope when writing to them.

Small tortoiseshell

Planning And Planting
The Butterfly Garden

Having understood the life-cycle of the butterfly, you are ready to start replacing the plants in your garden and generally remodelling it into a suitably 'wild' and fragrant habitat which will encourage and support them. In the course of doing this, however, you must be ready to turn your immaculate garden into a place which provides a habitat which is *ecologically more attractive* to these delightful insects.

Today the majority of gardeners have plots which are generally much smaller in extent and more meticulously cultivated than those of our predecessors. No weeds and very pretty, highly developed flowers is the conventional ideal. Unfortunately, this style of gardening does not appeal to butterflies

Red admiral

or bees, and many other insects in search of nourishment will also shun it. Gardeners have relied on modern technology to produce beautiful blooms, very few of which have enough scent to attract butterflies and moths into the garden. This situation is made worse by the fact that insects also have increasing difficulty in finding their natural food as more and more herbicides are used.

There was a time when one was able to walk through a cottage garden and breathe in the deep, heady fragrance of the flowers growing there. To walk down a country lane was to experience a world of beautiful scents from almost every hedgerow and wayside garden and the aroma drifting lazily on the still evening air encouraged nocturnal moths to seek their food, and to lay their eggs. All this has changed, but with a bit of planning and a reintroduction of some of the 'olde worlde' species of plants, we can do something to redress the balance, and at least provide a habitat which will be a better

Large white

place for wild creatures and offer some encouragement for butterflies and bees, as well as other animals. To encourage butterflies to stay, it is necessary that one's planting plan will not only provide a supply of food for the adults of the species, but also food plants where the female butterflies can lay their eggs and which will provide the necessary nourishment for hungry caterpillars when they hatch.

If you go to the trouble of providing suitable food plants for your butterflies you will probably want to know whether your garden is then a suitable place for them to visit. Unfortunately this is not necessarily so. It must also be remembered that not only will the geographical position of your home determine the species you attract, but the site of the garden is also instrumental. For example, you are unlikely to attract some butterflies if you live in the north of Scotland, because some species of butterfly are confined to southern counties of England. Similarly, if your garden is close to a wood and you are able to supply food for butterflies and food plants for caterpillars, then some wood-loving species *might* come into your garden. More detailed advice on the species of butterfly which may be found in various geographical locations throughout Britain should be checked in the specialist butterfly books which are recommended on page 94 of this publication.

Clouded yellow

77

If you are able to offer suitable food, but live in the centre of an urban environment, miles away from the nearest wood, the chances are that you will not have the species which normally spend their time in this particular habitat. However, if you are able to provide a sheltered garden which becomes warm and sunny then you are more likely to encourage butterflies to visit it than in a garden where, because it is open and windswept, the temperature remains low.

You will probably already have butterfly visitors, and, during a warm summer's day some of your bright flowers will become alive with the insects' dancing, vibrant wings. These butterflies have come

Meadow brown   Heath fritillary

to feed on the nectar. Just below the head of the butterfly there is a tongue, the proboscis, which, when the insect is not feeding, is rolled up tightly, rather like a watch spring. When the butterfly alights on what it assumes is a nectar-filled flower, the proboscis uncoils and it gently imbibes the sweet-tasting liquid. Having taken its fill, or found that there is not enough to satisfy its needs, the creature will leave.

Food plants for butterflies
When looking at food plants for butterflies and moths we have to consider them in two ways; there are the flowers which will provide the adults with their supply of nectar, and there are all the plants which will be a food source for the developing caterpillars. Let us look at the flowering species first.

There are two species of butterflies which, having spent the winter in hibernation, will need to search for a supply of nectar to replenish their energy before they start their arduous courtship ritual and egg-laying activity. The small tortoiseshell and the peacock will probably stay to breed if, on waking, they find a supply of nectar in our gardens. And with

the stinging nettles which we can plant we have at least the beginnings of a butterfly garden.

So we now have to decide which plant species will prove the most attractive to the insects in their search for nectar. When looking at flowering times we need to take the average into consideration. Cold weather will deter flowers from opening, and the position in the country affects growth. The particular time at which over-wintering butterflies will come out will depend on weather conditions and the part of the country in which you live. You may find it worthwhile keeping a diary of the butterfly species which makes the first appearance. To digress for a moment, if you are interested in 'firsts' then you will probably want to join the British Naturalists Association's Phenology Survey. Details of the address can be found at the end of this chapter. You will probably find that the small tortoiseshell is usually amongst the first and will be found, depending on the weather conditions, as early as March. Generally this species is joined by the brimstone, comma and peacock.

Once awake from their winter's sleep they will need food and we can provide it by growing plants of yester-year. Wallflowers are easy to grow, and these, together with Siberian wallflowers, the *Cherianthus*, will provide early nectar. Add to these low-growing plants like the perennial yellow alyssum, aubretia

Orange tip

and arabis, and you will attract plenty of butterflies to feed. Of the older cottage-garden species honesty is a lovely flower. Coming into bloom quite early in the spring, it will provide much needed food when, perhaps, the wild flowers in the surrounding countryside are not very great in numbers. Later in the year the seed pods of honesty provide a different sort of beauty in the garden. You could of course collect the seeds of all these plants, place them in labelled envelopes or polythene bags and sow them again where you want them to grow.

Closely related to the honesty is sweet rocket, which is an equally valuable flower for butterflies. Good strong plants will often reach a height of almost one metre (3 ft). The scent of this spring flower is so strong and the supply of nectar so great that while the plant is at its peak butterflies will be around for most of the day when the weather is fine. Its flowering times will depend on locality, but the main bulk of flowers appear, on average, towards the end of April and at the beginning of May. However, if you also want to encourage autumn

Small white

butterflies by having sweet rocket, seeds can be planted in the early days of spring. Although the flowers which appear on the sweet rocket in late summer/early autumn are smaller than the spring blooms, their attractive scent will still entice butterflies into the garden to feed.

Once you have established sweet rocket in your garden, it will become a permanent feature since it is a perennial. Nevertheless, if you want to do your own thing, and because the plant produces a good supply of seeds, you can collect these, and plant them in the following year to supplement the older plants, so injecting new life into the border. The seed pods are a tempting source of food for such hungry seed-eaters as the bullfinch, so you will need to protect some for your own use by covering some of the seed heads with muslin bags before they ripen.

I remember the pink thrift, so common along parts of the East Anglian and other coastlines, growing in my aunt's cottage garden in Norfolk. Although I watched butterflies coming to the plant, it wasn't until several years later that I realised the value of the plant as a source of food for butterflies.

In milder parts of the country stocks can survive the winter ready to bloom in spring, and these should be encouraged as a butterfly food. When selecting stocks make sure that you choose varieties which can be sown in the autumn, and which will flower in the following spring. In order to get the proboscis down to the nectar single flowers are easier for butterflies to manipulate and manouevre, and so should be planted in preference to double varieties.

Polyanthus and primroses – single-flowered species – will add a welcome touch of colour to the spring garden, and this will be increased as the brightly coloured flowers bring in the butterflies looking for food. It must be remembered that you will have butterflies for much of the year if you manage to plan your planting in such a way that you have a succession of flowers.

These are but a few of the many flowers which can be grown in order to attract butterflies. However, besides achieving this aim, they will also make an attractive display in the garden. In order to help you select other species, we give a table of those which are suitable on page 80.

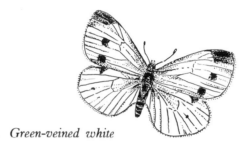

Green-veined white

Food plants for caterpillars

Apart from the small and large white butterflies which can feed exclusively on cultivated plant species, all other butterflies need wild plants as well as garden varieties. These wild plants – or weeds as some would call them – include the leaves of both trees and shrubs, as well as herbs. So to have any chance of encouraging these butterflies, you will have to give a large part of your garden over to weeds, and as someone who is concerned about the fate of our wildlife you will probably be prepared to do this. One of the ways of encouraging butterflies to visit the garden to breed is to encourage or even plant stinging nettles. The reason for this is quite simple. The most common species of butterflies recorded in most 'normal' gardens belong to the group known as the Vannessas. These are the bright red admirals, the peacocks and the small tortoiseshells. Although the cultivated plants are valuable to them as a source of food, they all need

SOME 'COTTAGE-GARDEN' PLANTS WHICH CAN BE GROWN TO ATTRACT AND SUPPORT BUTTERFLIES

The list below gives some idea of the plants which can be grown for butterflies. In reading other books and magazines you will probably be able to add to this list.

Buddleia	Lilac	Heliotrope
Michaelmas daisy	Aubretia	Allysum
Candytuft	Lavender	Bird's foot trefoil
Verbena	Scabious	Petunia
Catmint	Coltsfoot	Cornflower
Hemp agrimony	Golden rod	Primrose
Polyanthus	Honesty	Mignonette
Veronica	Lady's smock	Sweet rocket
Bramble	Ragged robin	Pinks
Sea thrift	Purple loosestrife	Ice plant
Sweet violet	Dame's violet	Stinging nettle
Blackthorn	Hawthorn	Honeysuckle
Hyssop	Marjoram	Thyme
Forget-me-not	Buckthorn	

SOME NIGHT-SCENTED PLANTS WHICH CAN BE GROWN TO ATTRACT AND SUPPORT MOTHS

Night scented catchfly	Bladder campion	Evening primrose
Petunia	Sweet rocket	Tobacco plant
Verbena	Honeysuckle	Soapwort
Alpine violet	Night scented stocks	Valerian
Californian primrose	Japanese honeysuckle	Everlasting pea
	White jasmine	

stinging nettles during the early stages of their life as caterpillars. The female will lay her eggs here, and here too the larvae will feed until they are ready to pupate.

Stinging nettles are not particularly fussy about their habitats; they will take root in almost any situation, in environments which are inhospitable to many more choosy species. You've probably seen the plant growing where ruined cottages stand, or by the hedge in your garden. Once established they, like many other weeds, are extremely tolerant and very difficult to get rid of. They grow particularly well where there is rubbish and rubble, or where soil has been thrown up. Yet once a 'disturbed' area has returned to 'normal' it is not unusual for the ubiquitous stinging nettle to disappear without trace. The reason for its departure is never very clear.

You can usually quickly encourage nettles to your garden, and at the same time save yourself muscle-power, simply by neglecting it, or part of it, since you will also want to experiment with some of the other ideas in this book, and so you won't want a complete nettlebed garden. To encourage butterflies to lay their eggs, the position of the nettles seems to be somewhat crucial.

Watching these butterflies in the wild I have seen them ignore vast tracts of nettles; a paradise one would have assumed for their egg-laying activities. What appear to be the criteria which will help establish a breeding colony? Probably to shade the eggs, and ultimately the newly-emerged caterpillars, from the wind, beds of nettles in slight depressions are generally favoured by the female. This, coupled with a sunny position, seems to present an ideal situation, and one can be almost certain of finding eggs and caterpillars if nettles located in a hollow are searched.

Depending on your garden, you will probably find that a stinging nettle patch close to an out-building – a garage or a shed – will be a suitable site. Adult peacocks and small tortoiseshells will seek out buildings in autumn for their winter hibernatory quarters. Having nettles near the building may encourage them to lay their eggs in the following spring when they re-emerge after their winter's sleep.

Red admirals also need nettles for egg-laying, so that the emerging caterpillars can feed on the plant. Although some red admirals may hibernate in this country very few survive this period of inactivity. So, although it is just possible that a few of these colourful insects may be ready to breed early in the year, the red admirals which we commonly see in our gardens are migrants which will have arrived from Africa. It is amazing that such fragile creatures should manage to complete such a strenuous journey, arriving here anytime from May to September, in the case of secondary migrations. Once here they will want to breed, an activity which will ensure the next generation will be ready to start the journey to Africa when our cooler autumn weather arrives.

By the time that the later red admirals arrive, the once fresh and succulent leaves of the stinging nettles are now much older and tougher, and not very pleasant for the caterpillars when they hatch. The same is sometimes true for the second brood of small tortoiseshells and peacocks. So in the wild garden it is necessary to tend the nettle bed. In order to provide young nettles when needed, a timetable of likely egg-laying, *etc.* is needed. Unless the plants are planned in this way, any female red admirals arriving in July will not lay their eggs on the tough, old leaves. However, they will eagerly seek out nettles which have been cut down earlier so that new, young shoots have grown up.

To cater for both red admirals and small tortoiseshells, it is a good idea, where the size of the nettle patch is reasonable, to cut down one third of the nettles in the early part of June, one third in early July, and the final third towards the middle of August. This last effort will ensure that there will be a supply of fresh nettles should any late arrivals visit your garden, and there is always a possibility that some will.

Those people who have a large expanse to devote to butterflies will undoubtedly want to include a small spinney or copse, or perhaps an orchard. There are some species of butterflies which will only breed in wooded areas, so that if we can provide this sort of habitat, we will be providing a valuable service for any butterflies which care to come and breed. This is of great importance since much of our deciduous woodland has already disappeared and much more is on the way out.

If you have a spinney, so much the better; if you intend to plant one, don't forget that it will take a while to become established. But even during the waiting period the area will prove valuable to wildlife – not only butterflies. If we have a spinney we will want to know what sort of butterflies we can expect to attract, and perhaps we might also be able to improve various aspects, which will increase the chances for butterflies to breed. In the table on page 71, you will see that we indicate the habitat which certain species prefer. If you check this, you will see that each butterfly has its own particular food species, and so you will need to make sure that these are available.

The butterfly bush
Butterflies have difficulty in feeding from many of the 'double' flowers which are found in modern gardens, and they have an aversion to red ones, preferring blue and purple flowering species. Perhaps the most popular, as its name suggests, is the 'butterfly bush' or buddleia. Apart from the well-known and widely grown purple variety, there is also a white one. Since it was brought to this country in 1896 it has proved invaluable as a food source for butterflies. Many escapee butterfly bushes are found in all sorts of strange places, including building sites, where it often manages to establish itself successfully. Depending on location, the first flowers will usually appear in July, and with careful management it is possible to prolong the flowering period. If the bush is pruned severely early in the year this will help. Continuous removal of the flower heads as they die off will encourage new ones to form later, and with this sort of attention there is no reason why the shrub should still not be bearing flowers as late as October. Such treatment will ensure that the insects out at this time will have a plentiful supply of food.

Encouraging And Keeping The Hive Or Honey Bee

Originally brought to this country from Asia to provide our ancestors with honey, the hive or honey bee has attained far greater importance in its ancillary role as pollinator. To recount the fascinating life history of this invaluable insect would take volumes, and some of the many books on the subject are listed at the end of this section (see page 94). The bee lives in a highly organized community where there is space only for workers and drones.

Each colony has three types of bee. These are the queen, the drones and the workers. There will only be one queen in the hive, and her prime function is to increase the population. At the height of her activities she will be laying as many as 2,000 eggs *every day* during the summer. The bees that emerge will be both drones and workers. The latter are very important in the hive, for the continuing success of the colony depends on their activities. In effect the workers are females, but without the ability to reproduce. Some will be making cells ready for the queen to lay fresh eggs; others will be out and about collecting pollen and nectar, and yet more will ensure that the young larvae are tended and cared for, which includes feeding, until they are ready to carry out the duties for themselves.

The drones are the males, and their *sole* purpose is to ensure that the queen is fertilized, so that a supply of young will continue to be born. Apart from this role, naturally a very important one, they have no other function to perform in the hive, and so, as the season of activity draws to a close, the drones are no longer needed. Indeed if they were allowed to stay they would consume vast quantities of food needed by the other bees. Almost callously, they are therefore pushed out of the hive and will perish. Although a colony of bees will last indefinitely, honey bees, like other species, are not noted for their longevity.

As pollinators bees are the natural friend of the gardener and farmer, but in areas where agriculture predominates and the use of poisonous sprays is an everyday occurrence, bees have suffered greatly, and their numbers have been reduced dramatically. How can the back garden wildlife enthusiast help? If he knows nothing about bees, then he can certainly get information and help from two sources. The first is the Beekeeper's Association of which there is a branch in most areas, the address usually being obtainable from your local reference library or Citizen's Advice Bureau. The other source of guidance is the International Bee Research Association which publishes some very interesting leaflets, and the research of which is extremely valuable and useful to everyone.

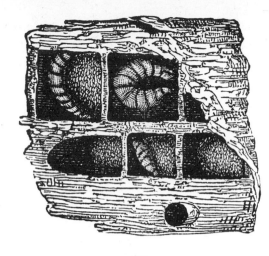

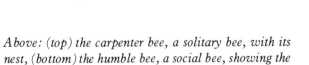

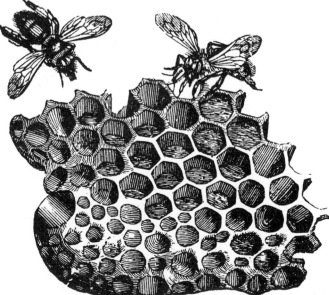

Above: (top) the carpenter bee, a solitary bee, with its nest, (bottom) the humble bee, a social bee, showing the jaws of male and female
Right: (top) the hive bee, showing worker, male and queen, (bottom) part of a honeycomb

Encouraging bees

To start with you can encourage bees to visit your garden. We suggest some of the trees and shrubs which you can plant which will help to provide a good supply of nectar for your bees in the table on page 85. Don't forget too, that the trees will provide valuable food for other species of wildlife as well. You will probably already be encouraging bees into your garden, because you will have

planted olde worlde flowers for your butterflies, and many of these will also attract bees. You may think the small patch of flowers you can provide will be ineffectual and perhaps hardly worth the effort, but you couldn't be more mistaken. If a large number of people did the same then the bees would soon have a good supply of flowers to visit, and these, hopefully, coupled with wild ones left untainted by poisonous sprays, will help ensure that the bee can continue in the vital role which it has to play. Perhaps one of the ways in which you can help the bees is to make sure that your garden is as poison-free as you can make it. As we have already mentioned, you can have a healthy, productive garden without using any poisons at all.

Perhaps you have often noticed bees visiting your garden, and seen them on your flowers. Probably these bees were seen at a distance, and at this range one bee looks much like another. Now that you have been converted you are, of course, much more interested in *all* the life in your garden, and you have probably already realized that

there are different species of bees, some of which are described later in this section. If you carry out a simple survey you will undoubtedly discover that some bees are found at some times of the day and not at others. There appear to be those which work to a given timetable.

Setting up your own hive

If you are very ambitious, you could even set up your own hive. You will be well advised to seek help from a beekeeping society and you could also consult the section on page 94, as many of the beekeeping suppliers will offer you a free leaflet or booklet to help get you started. The following notes are given to help you to decide whether to go ahead with the idea.

Many people assume that you have to live in the heart of the country to keep bees. This is not so, and many urban dwellers are realizing the advantage of keeping a beehive on a flat roof in the heart of our modern cities. In establishing a hive, not only will you be providing extra bees to carry out important work, but you should end up with your very own honey into the bargain.

As with keeping most 'animals' the initial cost and outlay is something which must be carefully considered. We have listed a number of firms at the end of this section, and you could write to them for their catalogues to find out the sort of equipment which they are offering; but you will probably decide to make some of your own and so save the initial cost. There are people who find it very difficult to produce anything, and for them the commercial supplier is probably a necessity.

Positioning your hive

The siting of the hive is a very important factor. Although you are dedicated to the 'conservation garden' your neighbours may not be. They are important, and must be considered. Position your hive away from any properties, including your own and your neighbour's. Avoid placing the hive where the bees come into contact with farm livestock. Although rarely, bees have been known to cause the death of animals after they have stung them, and a farmer wouldn't thank you if his prize heifer was killed in this way even though your bees might have been responsible for producing his bumper bean crop! The best position is to have the hive facing in a southerly direction when the

SOME FOOD PLANTS FOR BEES

Key: (P) Perennial
(A) Annual
(B) Biennial

Crocus (P)	Lemon balm (P)	Common sage (P)
Sage (P)	Thyme (P)	Creeping thyme (P)
Lesser celandine (P)	Borage (A)	Larkspur (A)
Cornflower (A)	Foxglove (B)	Common teasel (B)
Wallflower (B)	Narcissus (P)	Michaelmas diasy (P)
Delphinium (P)	Globe thistle (P)	Yellow archangel (P)
Lavender (P)	Virginian cowslip (P)	Catmint (P)
Lungwort (P)	Scabious (P)	Ice plant (P)
Alyssum (A)	Candytuft (A)	Snowy mespilus (P)
Barberry (P)	Japanese quince (P)	Rock rose (P)
Monkshood (P)	Lupin (P)	Lobelia (A)
Sea holly (P)	Bladder senna (P)	Dogwood (P)
Chives (P)	Bluebell (P)	Sea lavender (P)
Veronica spp (P)	Cotoneaster (P)	Nasturtium (A)
Clarkia (A)	French marigold (A)	Coreopsis (A)
Godetia (A)		

bees will have the advantage of the morning sun, although other directions may be necessary depending on your location, but *never* let it face north. To encourage the bees to set off at high level, it is an advantage to have a tall hedge in front of the hive.

Protective gear

Having decided on the site and possible position of your hive, the next important aspect before you buy the equipment, is to make sure that you are fully protected yourself. Although you will probably get used to being stung, and eventually take it in your stride, there are simple precautions which you should take. Suitable clothing is neces-sary when dealing with your insects. It is essential that *all* openings in normal clothes are secured — trouser legs, button holes, sleeves, *etc.* Further, it is *vital* to cover the face. You will probably have seen beekeepers with their faces draped with a net. Again, these can be bought, although they can be made without much bother. You will need about a metre to a metre and a half of Bretonne net, as well as a hat with a stiff brim. The latter is important because the net wants to protect your face, but not cling to it. Attached to the brim it will fall away from your face. The net is available in several colours, but black is preferable, as it makes for more distinct vision. Stitch the net so that it forms a bag — one end closed, one end open. The bag is then dropped over the hat, so that the sewn up end is over the top. You can secure it here if you wish with a few stitches. You will need a length of elastic, which has to be threaded through the back of the net, in such a way that the middle fits into the back of the neck. By bringing the ends through to the front – about 15 cm (6 in) apart – you can then take them under the armpits and secure them at the back. This offers an effective 'cage' against any inquisitive bees!

Purchasing a hive

Your greatest initial outlay will be for the purchase of a hive. To buy a new one will set you back about £30 to £35. There are alternatives of course. You could put a *wants* advertisement in the local press or in one of the beekeeping magazines. The other way of combatting the high cost of a manufactured hive, especially if you are a handyman, is to make one of your own. You can find details in *Beehives (Bulletin 144)*, published by Her Majesty's Stationery Office, and available from your library, or by ordering it through your local bookshop. The hive that you purchase will include all the necessary 'furniture' so that you can start right away.

Buying a colony

Having bought or made a suitable hive the only thing which is now missing is the bees! You may have been able to purchase a hive which already contains a colony. If it does not contain a colony of bees you will need to purchase one separately. You should examine the pages of one of the beekeeping magazines for further sources. Naturally the price fluctuates from year to year, but you can expect to pay about £15 to £20 for a large colony – around 25,000 workers. This will include all the necessary members. Smaller colonies are obviously cheaper. If need be queens can be purchased separately. Now you have the bees, hive and veil, you are ready to start, except that you will need a smoker. Smoke encourages the bees to become very active, with the result that they take their fill of honey. Then, like a man after a good meal they are much more lethargic and docile.

Making a bee-proof window

We have already suggested that you keep your hive away from residences. The same also holds good for your beekeeping equipment. If you have a garden shed you can use this: all you will need to do is to carry out one or two modifications. Check the building for security, ensuring that there are no likely entrance holes for the bees. However, from time to time, particularly when the door is open, they will make their way into the shed. Although you may feel you have a 'bee-proof' shed, there are likely to be some means of entry, which are not usually discovered, so that when the bee wants to leave again, it will have difficulty. You will need to modify your window, so that the bees will be able to get out.

Remove the original glass and if you have suitable tools, cut it so that it is about 15 mm (0.6 in) shorter

than the depth of the window. Refix the glass, leaving a gap at the bottom. You will need another piece of glass, the same width as the window, but only about 5 cm (2 in) deep. This is placed in position, some 15-20 mm (0.6 – 0.8 in) from the first

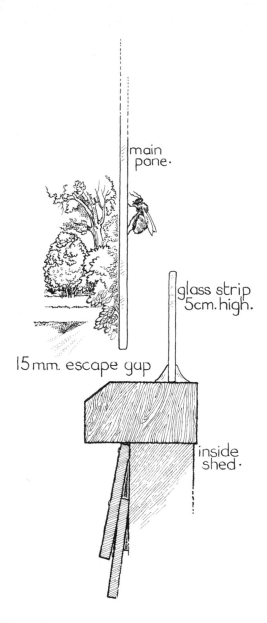

main pane.

glass strip 5cm. high.

15mm. escape gap

inside shed.

Making a bee-proof window as described in the text

piece. If you have more than one window in your shed you will either need to cover the rest so that they are lightproof, or repeat the operation suggested here on all the windows. If bees do get into the shed they will have a means of escape. Their immediate reaction will be to make for the light. As they climb up the inside glass, when they reach the top, they will drop down, and they can escape. Some people prefer to fix the small piece of glass on the outside, although rain may then be able to come in.

Winter feeding

During the colder months of the year when the bees are unable to collect their food they must still be fed. It is important that a sufficient supply of honey is left in the hive during this period. It is usual for each full hive to produce around 18 kg (40 lb) of honey each year, but of this amount 13.6 kg (30 lb) must be left for the winter period. Some apiarists take out 15.9 kg (35 lb) and only replace 6.8 kg (15 lb), making up the rest with a mixture of sugar and water, but this isn't really a suitable alternative to their natural food – the honey.

Swarming

Whether or not you have a hive in your garden you could become the target for the attention of a swarm of bees. If a colony is particularly well established and prolific then the numbers might increase to such an extent that life is becoming very difficult inside the hive and at some pre-arranged signal a large number of bees will suddenly leave. There will be one queen and she will be accompanied by a large number of workers. Generally they will come to rest on a tree not far from the hive which they have just left. A few of the workers will leave the rest of the swarm to search for a suitable new home. A local beekeeper may well have been informed of the swarm and will soon collect them and take them to an empty hive. A swarm of bees on the loose, as it were, tends to cause the observer to panic. But the insects' sole purpose at this time is to find somewhere which will form the basis for a new home. Furthermore, presumably as a precaution against not finding food or settling for a while, the bees have taken more than their fill of honey and are so bloated that the chances are they are even incapable of using their only defence mechanism – the sting! In general bees are placid insects. Their sting is a means of defence and if upset or annoyed they have very little option but to use it. If you intend to set up a hive in your garden then one of the things which you should avoid doing is standing in the flight path to and from the entrance to the hive.

It is likely that once a hive becomes established from a swarm, it will be queenless, but within a short time the workers will have made sure that some of the developing grubs are well looked after, and this extra attention causes them to mature with the result that they become the new queens. Only one will take charge of the colony, but the mechanism of the 'selection process' is not really understood.

The Bumble Bee

The bumble bee is a large, hairy insect with a deep drone, whose plump body and seemingly lethargic flight as it moves with what seems to be a great deal of effort from one flower to the next, give the garden an air of tranquillity. This difficulty in flight is due to the extra weight which these insects carry. Careful examination usually shows that many bumble bees are transporting large numbers of 'visitors'. These are very small, brown-coloured mites which generally seem to cling to the bee's thorax. With large numbers of these mites to carry, flight seems to be less efficient and sometimes the bees experience difficulty when taking-off. In spite of the apparent awkwardness which these guests cause, as far as is known they do the bee no harm. Once the bee's nest is completed, these tiny parasites will seek shelter and food there, and she will soon be free of most of them.

Bumble bees, unlike some of the other species which are discussed later, are more 'sociable'. They do not live solitary lives, but settle for a communal nest. Although the nest is miniscule in comparison with the hive bee colony, nevertheless a communal spirit exists. Each nest will have a number of individuals not only living, but also working together. During the winter the young queen bumble bees will have been hibernating in some quiet, protected part of the garden until the warm spring weather stirs them into activity. When they first awake they will seek out nourishment before they turn their attention to the all-important task of building the nest.

Encouraging the bumble bee to nest
In looking for a natural place for this the bumble bee will generally search out disused nests, once occupied by some of our other garden inhabitants, the mice. It is fascinating that the bumble bee should specifically search out the old nests of mice, and although you may not have mice resident, you can create an artificial, but nevertheless, life-like site for your bees.

It is possible that you might have difficulty in searching out this nest material. One way of building up a supply is to trap mice in your garden and keep them in captivity, providing them with bedding material which, within a short time, will become 'mousey' and should serve the purpose. Alternatively you may know of friends who keep mice, and you could ask them for some old bedding from a nest. Many schools have animal rooms and keep mice and you could perhaps get some material from this source.

Having located a source of supply you will need to select a suitable site and then dig a hole big enough to take a medium-sized flower pot. The queens prefer to nest in undisturbed areas where they can carry on their task unmolested. Unkempt areas of rough grass are suitable, with sunny sites taking preference over shady ones. If you have a hedgebank with a south-facing aspect, the chances are some bumble bees will take up residence.

Having selected a suitable site for the artificial nest and dug a hole to take it, partially fill the pot with your nesting material, leaving a gap at the base. Earthenware pots are better than the plastic variety although these are sometimes difficult to obtain. With the drain-hole at the top, lower the pot into the ground, so that the base (now at the top) is level with the soil surface. Using five stones – four small ones and a larger one – you will be able to provide a cover for the pot, so that water will not get in. Place the four smaller stones around the edge of the pot and rest the large stone on top. There will then be a gap to allow the bees to crawl underneath and into the hole.

Artificial Homes for Solitary Bees

In addition to the social insects which live in permanent colonies like the hive bees, or in seasonal colonies like the bumble bee, there are solitary bees and some wasps which have a need for holes in which to nest. As a youngster I lived in an old house and I used to watch fascinated as bees went into holes in the wall between the building blocks. Many species prefer to excavate their own nests, although if pre-drilled ones are available

Foxglove visited by bumble bees

sawn up logs. Pieces of log about 30 cm (12 in) in length are suitable. Drill holes around the outside of the log to the depth of your drill bit. Once you have done this find suitable spots in your garden to place them in. The insects prefer warm quarters, but beware of sun traps. In corners of gardens and particularly patios where the morning sun shines on them, the temperature can soon rise dramatically. Obviously the adults can quickly move out, but if larvac are inside the excessive heat could kill them.

One of the drawbacks is that you cannot see the exciting work which goes on inside the nests. It is possible to buy glass tubing of various diameters, perhaps from a local chemist or probably from a supplier. If you plug one end of a piece of glass with some plasticine and then push the glass tube carefully into one of the holes, then solitary insects *might* use this particular tube, and once you have seen that one of these glass tubes is occupied you can watch progress. Many of the solitary species make small clay cells and the female lays a single egg in each of these. Cells are placed one in front of the other inside the tube.

Other artificial homes

You may find that other ideas are useful if tubes can be provided. One which has been used successfully is to fill a hanging basket with dried grass, and then push some hollow tubes into this. Hollowed-out bamboo canes, cut to a length of about 15 cm (6 in) will serve the purpose. You could also make your own by using lengths of elder stems. These have soft pith which can be removed. You could make your own basket using chicken wire, which is obviously cheaper, and suspend it in a suitable position – you probably won't want it in the porch, as it will be difficult to observe.

We mentioned earlier that there are those species which prefer to make their own nests. Known as digger wasps, for obvious reasons, they will be happy to excavate their own tunnels. If you can provide some decaying wood – logs are suitable – they may take advantage of your hospitality.

You might also be able to get hold of a ventilation block – the sort that one puts in the outside walls of houses – from a local builder. This serves a similar purpose to the holes in the logs. However, because

they may use them, so it is worth providing some suitable places.

If making tunnels for themselves these insects will dig them out to fit their own bodies and obviously, as insect body sizes vary, it is necessary to provide different sized holes. It is difficult to give exact measurements as you can imagine, but drilled holes should vary in size from about 10 mm (0.4 in) for the largest, down to about 3 mm (0.1 in) for the thinnest. Wood is by far the easiest material to drill. Having sorted out the necessary drill bits, you can drill a number of holes in some the holes are often square, and too large for the insects to use, they may make adaptations. Alternatively, you could make your own alterations by adding either hollowed out elder stems or bamboo canes. If you have a wall somewhere, or a raised pond, you might replace one or two bricks with air ventilation blocks. Alternatively you could put up a small wall in one area of your garden. This will also be advantageous to other species, because there are many invertebrates which will seek out the shelter and protection provided by the bricks.

Another idea is to make up bundles of hollow tubes – bamboo or elder – and place these under a window ledge of an outhouse. The bundles can be held together with elastic bands or waterproof tape. Straws make a suitable alternative, especially if plastic ones are available. The disadvantage with using straws under a window ledge is that run-off water from the ledge trickles underneath and may ruin them.

If you want to find out as much as possible about bee visitors to your garden, some further reading is suggested at the end of this section on page 94. However, some notes on various kinds of solitary bees are given here to whet your appetite.

The leaf-cutter bee
One of the most fascinating species of solitary bee is the upholsterer or leaf-cutter bee, although it is only found in the southern counties of the British Isles. Not unlike the honey bee in general size, it has one distinguishing feature, its large jaws, which have obviously evolved because of the neat 'cutting-out' job they perform. Using its sharp jaws the bee quickly cuts out a circular piece of leaf. Particularly partial to roses, once she has removed a piece she curls it up and, with the aid of her legs, holds it under her abdomen. Once back at her nest she uses the piece of leaf to line out a cylinder-shaped cell. She will lay one egg in each of the cells, having also ensured that there is a supply of honey for food when the larvae hatch out. Once all the eggs have been laid with a suitable food source at hand the female has completed her job and will die. Her eggs hatch out, but they do not develop into adults immediately and remain as pupae until the following spring.

The cuckoo bee
Cuckoo bees are aptly named, for, like the birds after whom they are named, they are parasitic.

Did the cuckoo bee evolve its parasitic habits because it had no pollen baskets on the hind legs, and would have difficulty feeding its young, or did the pollen baskets disappear as it adapted a parasitic mode of life?

Once the cuckoo bee has emerged from its winter hibernation, it seeks out the nests of bumble bees. It appears that the female cuckoo bee is able to locate the bumble bee's nest by a characteristic smell which is given off. This is due to the fact that the bumble bee chooses old mice nests for its new home – and mice certainly have an unmistakeable smell!

Once the cuckoo bee has located a home the life of the bumble bee will soon come to an end. The queen bumble bee will already have laid her eggs which will have hatched into workers. The cuckoo bee will lay her own eggs and shortly afterwards she will kill the queen which has no further part to play in the new order of things. When they emerge, the bumble bee workers will take over the care of the cuckoo bee larvae. The cuckoo bees will turn out to be mainly drones – males – together with some females. No workers are necessary as the bumble bees provide these. The cuckoo bee drones will provide the necessary mates for the queens. When the time comes for hibernation, the queens, already mated so that they can lay their eggs in the following spring, will seek out suitable shelters.

Not long after the bumble bees emerge from hibernation the cuckoo bees are likely to be found in the garden as well. Their hibernation lasts longer than that of the bumble bees because they have to wait until the latter have prepared their nests and laid their eggs. You may have difficulty in distinguishing the cuckoo bees from the bumble species because the former often resemble their host, in order to 'get away' with their parasitic habits.

The mining bees
Although the hive or honey bee, is probably the earliest to appear in the garden, it will soon be accompanied by other early visitors. They are members of the genus *Andrena* and they do not usually have common names, although they are often known collectively as mining bees. These bees have acquired the 'mining' tag because they live in the ground. If you have a non-concrete

garden path you will quite likely see them as they flit in and out of their holes. The chances are that you will also see larger numbers quartering the ground as they search out the females. When they first appear the only flowers available are generally dandelions, and they will visit these regularly in their search for a supply of food. As the early aubretia and other spring flowers come into bloom these bees can be seen buzzing almost non-stop around the scented plants.

All species of *Andrena* live solitary lives. Each queen has her own burrow where she will lay her eggs. Having provided a supply of food for the larvae when they hatch, she will no longer have any interest in the nest, and in effect, abandons it. As the eggs hatch they will feed on the food which the female provided. Even at this stage they are solitary by nature, each going its own way. Once they are fully grown they will leave the nest. It is

likely that some members of the early brood will mate and produce offspring of their own. Most, however, will not be ready to breed until the following spring.

The red osmia bee
As the apple blossom starts to break open from its protective bud in April and May, the red osmia bees will visit the garden to take their fill and also unwittingly to assist with pollination. This is one of the species which might take up residence in the hollow tubes which we have suggested you could provide in your garden. There are other places where they will also build their nests, including holes in both tree trunks and walls. If you have a wooden shed with overlapping panels, you might find a red osmia bee disappearing behind these, and here it could make its nest and lay its eggs.

Wasps

Many insects which have a general resemblance to wasps are labelled in the same way. All the wasps which visit the garden are likely to fall into one of two groups: they will either be true wasps or digger wasps. The many different species which make up the group known as true wasps include some which live a solitary life, as well as social insects. Among the diggers there are those which also excavate holes, much like the digger bees. Some are so good at making their homes in old walls that they are quite rightly known as mason wasps. Their other favourite haunt is in banks. In both instances the digging of the nest can be a protracted affair. The mud – or sand in the case of mortar – has to be softened before it can be removed. This the wasps do with water. One of the strangest activities of some species, but as you will see, a very sensible one, is the building of small pipes at the end of the nest. In walls the insects use the sand which they have excavated to make these tubes. By bending them they are able to have protection, not only from rain, but also from would-be attackers. Once the mason wasp has made its nest to its own specifications, it will lay its eggs, and then usually search out caterpillars with which to feed the larvae when they hatch. Using its effective sting it does not kill, but simply paralyses its victims. This is particularly important because it could be some time before the larvae get round to consuming their food, by which time it could have putrefied had it been dead.

Of the six species of social wasps, not counting the less common hornet, only two are likely to be encountered in the garden. Known as the German wasp and the common wasp, there is very little to distinguish one from the other, except the facial features. On the face of the German wasp there are small black spots in contrast to what is considered to be an anchor-shape on the common wasp. This is not always an accurate method, because it is not unusual for the anchor in the common wasp to be replaced by small dots, not unlike those on the German wasp!

Like the bees, the queen wasp produces a nest in spring. There are large numbers of queens, and as soon as the weather is warm enough, they will be out searching for suitable nest sites. It has been estimated that only about one in every hundred queens manages this first and most important task of its arduous life. Although the first wasps will usually stir in April, when there have been warmer spells in earlier months these insects, along with many others, have been lured out. Quite often these premature, inflated temperatures are followed by frosts and the wasps are killed.

There is much in common between the social wasps and their relatives, the social bees. As with the bees, the only wasps to come through the winter are the pregnant queens. Having selected a suitable over-wintering site, often in outbuildings or under the roof, even in houses, here the queen will rest for several months. Having need for nourishment when they wake up in the spring, the queens can be seen seeking nectar from various flowers. However, the all-important task is to produce a nest as soon as possible, and quickly they turn their attention to this project. Most wasps make their nests below ground level. Suitable sites will be eagerly sought, and a disused small rodent's nest might be discovered and used. The nest is made from wood. The female's jaws are extremely strong and she will soon tear off a thin strip of wood. Before flying back to her nest site she will curl up the wood, and carry it in her jaws. Once back to the building operation she mixes the wood with some saliva which softens it. Carefully she spreads it out, so that it forms a thin sheet. Working continuously and deeply engrossed in her task, within a few days she will have the foundations of a nest. Perhaps 'foundations' isn't quite the correct term, because she starts by making a roof. Attached to the underside of this she makes a few cells, where she will lay her eggs. This interruption to the building programme is short-lived and she continues until the larvae hatch.

Now her task is two-fold; the larvae demand a great deal of attention, and she has to counter this with her building commitment. As the larvae grow they win, and the building operation decreases. The main source of nourishment comes from caterpillars and other insects, which the queen brings. Before she offers them to her offspring, she will make sure that they are more palatable by chewing them up! Whereas solitary wasps sting their prey to immobilize them the social wasps bite theirs. Some five weeks after the first eggs were laid the newly-hatched wasp workers relieve the queen of much of her work, which leaves her free to concentrate on her egg-laying activities. The newly hatched workers will help to increase the size of the nest ready for

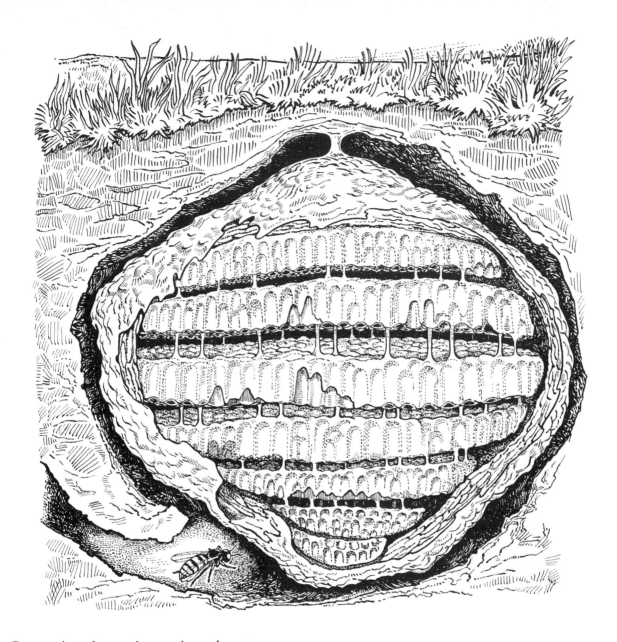

Cross-section of an underground wasp's nest

new eggs and workers and will also feed the larvae as they hatch out.

Unlike their relatives, the digger wasps are all solitary species. In the garden, the females will usually search out rotten and/or decaying wood which they find easy to get into. Some of these wasps will not even bother to build their own nests. Instead they will seek out hollow tubes. Canes used in the garden are quite popular. If you have already made this provision, you are likely to attract these insects. In areas where thatched roofs feature in the landscape, the smaller species are likely to seek out the narrow tubes of the reed used on the roof. A few eggs will be laid in single cells. These wasps too will paralyse various grubs and insects, which they will carry back to their nests for food for their offspring.

Further Useful Information

Organizations and Societies

Amateur Entomologists Society, 18 Golf Close, Stanmore, Middlesex

The Society, founded in 1935, promotes the study of entomology, particularly among amateurs and the younger generation. Produces a wide range of publications on insects.

British Bee-keepers Association, 55 Chipstead Lane, Riverhead, Sevenoaks, Kent

British Butterfly Conservation Society, Tudor House, Quorn, Leics, LE12 8AD

The Society is concerned about the plight of Britain's declining butterfly population. A leaflet, *Butterfly Conservation*, is available from the Secretary, who will also be pleased to give further information about membership. Produces a very interesting and informative Newsletter.

International Bee Research Association, Hill House, Gerrards Cross, Bucks, SL9 0NR

This Association is concerned with all aspects of bees, and has a valuable and extensive library of books which is available to members. Produces pamphlets, books, *etc*. It also provides the only central source of scientific information about bees and beekeeping, together with the very important related aspect of insect pollination.

Royal Entomological Society of London, 41 Queen's Gate, London SW7 5HU

The objects of the Society are the improvement and diffusion of entomological science. Publications include a series of books which help with the identification of insects.

Periodicals and Magazines

Bee World, International Bee Research Association, Hill House, Gerrards Cross, Bucks, SL9 0NR

Published four times a year, and is available to members.

Books, Leaflets, etc

Burton, J., *The Oxford Book of Insects,* Oxford University Press

Butler, C. G., *The World of the Honeybee,* Collins

Chinery, M., *Insects of Britain and Northern Europe,* Collins

Joint Committee for the Conservation of British Insects, *Code for Insect Collecting*

Ford, E. B., *Butterflies,* Collins

Ford, E. B., *Moths,* Collins

Ford, R. L. E., *Studying Insects — A Practical Guide,* Warne

Free, J. B. & Butler C.G., *Bumblebees,* Collins

Goodden, R., *British Butterflies — A Field Guide,* David & Charles

HMSO, *Beehives,* Bulletin 144

Higgins, L. & Riley, N., *Butterflies of Britain and Europe,* Collins

Howarth, T. G., *Colour Identification Guide to British Butterflies,* Warne

Hyde, G. E., *British Butterflies* (Books 1 & 2), Jarrold

Hyde, G. E., *British Moths* (Books 1, 2, 3 & 4), Jarrold

Hyde, G. E., *Butterflies* (Town & Country Series), Almark

Imms, A. D., *Insect Natural History,* Collins

International Bee Research Association, *Trees and Shrubs Valuable to Bees*

More, Daphne, *The Bee Book,* David & Charles

Moths and Butterflies of Great Britain and Ireland (11 volumes), Curwen Books

Newman, H. L., *Create a Butterfly Garden,* John Baker (also available as a paperback)

Newman, H. L., *Looking at Butterflies,* Collins

Newman, H. L., *The Complete British Butterflies in Colour,* Ebury Press/Michael Joseph

Proctor, M. & Yeo, R., *Pollination of Flowers,* Collins

Urquhart, J., *Keeping Honey Bees,* David & Charles Penny Pincher

Scott, W., *Backyard Beekeeping,* Prism Press

South, R. & Howarth T. G., *South's British Butterflies,* Warne

South, R., *Moths of the British Isles,* Warne

Stokoe, W. J., *Observer's Book of Butterflies,* Warne

Tweedie, M., *Insect Life* (Countryside Books) Collins

Tweedie, M., *Pleasure from Insects,* David & Charles

Von Frisch, K., *The Dancing Bees,* Metheun

Equipment and Supplies

Baxter, R. N., 16 Bective Road, London E7 0DP:
Living stages of British moths

Birdwood Apaiaries, Hawkers Lane, Wells, Somerset:
Bee equipment

The Butterfly Farm Ltd, Bilsington, Ashford, Kent:

All stages of butterflies and equipment

Lee, Robert (Bee Supplies) Ltd., Beehive Works, George Street, Uxbridge, Middlesex:

Bee equipment

Taylor, E. H. Ltd, Beehive Works, Welwyn, Herts, A16 0AZ:

Bee hives, wax foundations and all appliances. A detailed catalogue, together with a very useful booklet, *How to Begin Beekeeping*, available on request. The firm also supplies swarms of bees

Thorne, E.H. (Beehives) Ltd, Beehive Works, Wragby, Lincoln, LN3 5IA:

Beginners' leaflet and full catalogue available on request. Has wide range of equipment and supplies

Watkins & Doncaster, Four Throws, Hawkhurst, Kent:

Sells various equipment for the amateur naturalist, including entomological supplies

Worldwide Butterflies Ltd, Compton House, Sherborne, Dorset, DT9 4QN:

Livestock, equipment, books, *etc*. Has a wide range – catalogue available regularly via a mailing list for which a small charge is made

Miscellaneous
British Naturalists Association Phenology Survey

The idea is that people in various parts of the country record the first sightings of, for example, the various butterflies. Further details from the Hon. Secretary (see entry in Chapter 1, page 7).

THE
GARDEN POND
AND ITS
INHABITANTS

The illustration shows (anti-clockwise, from top left): yellow flag irises, yellow water lilies (in pond, at the back), water skater, common frog, frogbit (plant on the pond surface), various pond weeds oxygenating the water, great diving beetle, great ramshorn snail, smooth newt, great pond snail, white water lillies, water plantain, water crowfoot (small leaves on the pond surface), monkey flower (centre back of the pond)

No one will deny that as more and more ponds disappear, there is a need to replace the natural form with artificial ones. The Save the Village Pond Campaign has done much to help renovate and rejuvenate many of the old ponds which, for various reasons, have fallen into decay and disrepair.

Although we stress that one of the greatest threats to the well-being of amphibians supported by these ponds is man, there is another factor which must be remembered. Over a period of years a pond will naturally become clogged with silt and water plants, so that it will eventually revert to dry land. This process begins when the reeds and rushes which grow by the water's edge start to encroach into the water. Soil will gradually collect around the roots of these plants and the once open water gradually will become marshy and eventually dry out altogether. With the upsurge of interest in village and other ponds, there is again an increase in these freshwater habitats. This is good news, and with the addition of garden ponds, provided that they act as a reservoir for wildlife, the balance might be redressed to some extent in the years to come.

Making An Artificial Pond

At one time the thought of an artificial pond in the garden filled the would-be enthusiast with despair as he contemplated the work involved. Not only did the site have to be excavated, but concrete had to be mixed and left to be cured once the pond had been finished. Today it is still possible to go about things in the same way, but fortunately, with modern technology and the advent of polythene and rubber sheeting, as well as the preformed garden pond, life can be made much easier, and a pond in the garden can be a reality for almost anybody who wants one.

Safety considerations
Assuming, therefore, that you have decided you have room in your garden for a pond, you will need to give some thought to for example, the size and shape and material to be used. An important point which needs to be made at this juncture is the question of safety as far as children are concerned.

Tragedies have occurred when young children have drowned in garden ponds. Thus it is not advisable to have a pond which is likely to cause unnecessary worry, not only as far as your own children are concerned, but those of friends and neighbours as well. There are ways of making a pond safe. A strong childproof fence all the way round will go some way to solving this problem, but this does, of course, detract from the amenity value. Mesh can be used to cover the surface of the pond. Again, this does have disadvantages in that it prevents birds from getting at the water – perhaps in some instances this might not be a bad thing!

If large numbers of water plants are placed around the edge of pools these will usually deter children – but there are always exceptions. If you think that there is any possibility of danger *DON'T* have a pool. Your children won't always be young and it is something which you can add to your wildlife garden when they have grown to the 'out of danger' age.

Choosing and siting your pool
Once you have decided that a garden pool will make a useful addition to your garden as a wildlife sanctuary, the following will need to be considered. The best time of the year to start is early spring, and if you begin operations in March, you will have the following months to establish your new habitat. Whatever the size of your pond you should always endeavour to have at least two levels. Some species, both plants and animals, enjoy the type of aquatic habitat provided by deep water. Shallower areas are important for other species. You may also decide to have a marshy or very shallow area, which will attract birds and be a valuable asset.

There are several factors which you need to take into consideration when deciding on the pond's position. If it is possible one should avoid trees. Apart from the fact that they may cause dense shade which will deter plant growth to some extent, they are also likely to be a problem in the autumn when the leaves fall, and during the whole year if they are conifers. Most experts suggest that it is a good idea to have a pool in the open so that during the summer it will be in sunlight for about eight hours a day. Such a situation might affect shade-loving animals, but this can be remedied by having large plants, like water lilies, which offer a great deal of shelter.

The size of your pool is obviously going to be determined by the size of your garden – at least this is so where smaller patches are concerned. There are other factors to be taken into consideration as well. The material which you decide to use for lining the pond will also to some extent affect the design and shape. You have a choice of several materials, and the merits of each are discussed later.

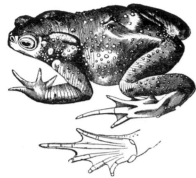

The common toad

Concrete

Until the advent of modern materials, concrete was the only material with which ponds could be lined, except for puddled clay in some instances. The new materials have helped tremendously and made the garden pond a reality for many more people. But what are the problems with concrete? Apart from the obvious one of toil and sweat, concrete will often crack. Freezing temperatures, especially when prolonged, often bring about cracking, and with excessive dry periods the soil may dry out below the pond with cracking again a result. Although cracks can be repaired, it does mean that the pond has to be emptied, which is an inconvenience, and once the pond is re-filled it has to become re-established, which may take some time.

In spite of the disadvantages given above there are people who still prefer to use concrete as a lining, and when this is the case the following factors must be taken into consideration. Firstly, the size of the final pond will be much smaller once the concrete has been put in. In other words it is necessary to allow for the depth of concrete. The advantage of a concrete pond is that it can be made larger, although some of the stronger lining materials do not limit the size of the structure.

Another factor is the importance of getting the concrete mixture right. When the shape of the hole has been dug, allowing for the thickness of concrete, the bottom needs to have 5 cm (2 in) of rubble – hard core – as a firm base for the concrete. The ratio of materials for the concrete mix itself is:

> 3 parts of coarse aggregate
> 2 parts sand
> 1 part cement

Having measured these out, mix them well with water to form a thick consistency. This achieved, the bottom and sides need to be covered with a thickness of about 10 cm (4 in). Obviously the walls will need to be held in position, and shuttering should be used. This can be removed when the concrete is dry. The final rendering coat has now to be applied. This should be made by mixing:

> 3 parts sharp sand
> 1 part cement

obtainable from a builders' merchant.

The concrete needs to be waterproofed, and a waterproofing powder needs to be added to the cement. These ingredients are mixed with water to a smooth consistency and applied to the bottom and sides of the dry pond. The waterproofing

The female eider duck

powder will only work if it is mixed evenly and the coating applied in a uniform layer.

One of the dangers of concrete is that it contains free lime: this affects both plants and animals, so it is necessary to get rid of it by filling the pond with water, which needs to be left for a week. This water is emptied and the pond re-filled. Again this is left for a week, before being emptied.

99

A third filling is necessary and this will usually ensure that any remaining lime will be removed from the concrete. The pool will now be ready for stocking as described later. As an alternative method of freeing the pool from lime, the whole surface of the fully-lined pool could be treated with a neutralizing material: Silglaze is one which is frequently used for this purpose.

Materials other than concrete

The materials described below are generally considered more versatile and of greater durability than concrete. Preformed glass fibre comes in a variety of shapes although the limiting factor is size. When compared to sheet lining materials, preformed designs are very expensive. However, it has many properties which make it excellent for pools, including the fact that it is very strong, and although it has a tendency to 'rigidity' it will give with the weight of water. In general, preformed glass-fibre pools are shaped with shallow and deep areas, and one of the problems is in digging the correctly shaped hole. As with a concrete pool the first task is to dig the hole to take the shape: the hole needs to be slightly larger than the size of the pool. When the unit has been placed in position, sand should be used to fill the gaps under and around it. The problems which arise when using concrete do not arise with the glass-fibre ponds, and they can be stocked almost immediately.

Using one of the plastics or butyl rubber as a pool liner has great advantages. Of the artificial materials available (pvc, butyl rubber and plastolene) there do not appear to be any advantages or disadvantages in selecting a particular type, but it is worth investigating the costs, as these might vary from area to area. The great advantage with these materials is that they are very adaptable. None really has any size limits, and the shape of the pool can be as varied as the designer wishes. At first sight one might be forgiven for thinking that pvc is nothing more than polythene, but closer examination and comparison will reveal that it has greater elasticity, and is also thicker. Butyl rubber has similar properties to pvc, although it probably exhibits slightly greater strength characteristics. Both butyl rubber and pvc will be found under various trade names. Pvc has been welded on both sides of a layer of terylene and probably has more strength than the other two. Plastolene is the third material which can be considered.

As with either concrete or fibre glass, the size of the pool must be decided before excavations begin. You could be adventurous with your shape, but try to remember that different levels are important too. Once you have decided on the shape of your pool, you will need to work out the area of material required to fill the hole! When working out the length of the liner, you will need to add twice the greatest depth to the overall length. The width required will be twice the greatest depth added to the overall width.

The hardest task is again the excavation of the hole. The easiest way to ensure that the shape is correct is to mark it out with string. If the area is covered with turf, this should be removed with either an edging tool or a spade. The hole can now be excavated, not forgetting to leave shelves and a marginal area for a bird-bathing place. The sides should be sloping, and an angle of 20° is probably a useful one to aim for.

Once the soil has been removed, it is a good idea to cover the bottom of the hole with fine sand, making sure that there are no sharp pieces. Although the suggested lining materials are extremely tough, sharp stones will quite easily puncture them, especially once the water has been added. Peat is an alternative material which can be used for the lining, and in case neither of these is available, there is an equally good, free source – newspaper, which can be placed on the bottom.

This operation completed, the pool liner can now be placed in position, ensuring that it is touching the bottom and sides. Bricks or stones should be used to hold the liner on the surface around the pool, and then the pool can be filled. Using a hose, try and ensure that the water trickles into the middle. The weight will soon 'spread' the liner, and mould it to the shape of the excavation. Water should be added until it is about 2.5 cm (1 in) or so from the top. Although the edge would do as it is, most people will want to make a neat job of the pool. The polythene should be trimmed, leaving about 15–20 cm (6–8 in) all round. The edge can be hidden under the turf, or covered with paving slabs or crazy paving. Crazy paving is quite useful, since gaps can be filled with soil, rather than cement, and plants will colonize these areas. By allowing the paving stones to hang slightly over the edge of the pond a neat effect can

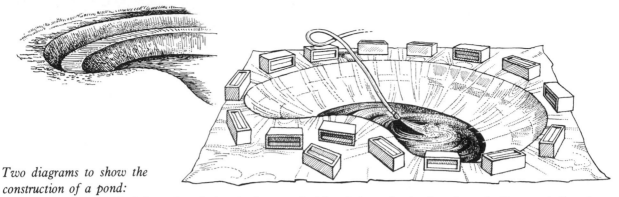

Two diagrams to show the construction of a pond:
1   The hole is dug with shelves, then the interior is covered with polythene sheeting (or other suitable material) and held in place with bricks. It is then filled with water.

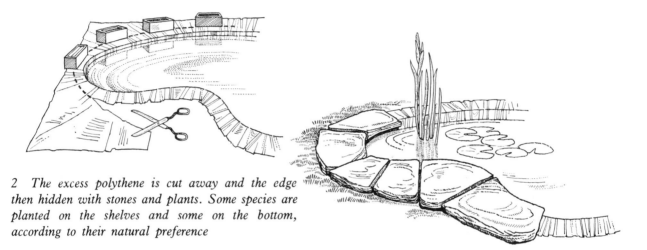

2   The excess polythene is cut away and the edge then hidden with stones and plants. Some species are planted on the shelves and some on the bottom, according to their natural preference

be obtained. Be careful here, because if they hang too far over this could be dangerous if someone stands too close to the edge. Many animals also, such as hedgehogs and various rodents, could topple into the water, if they happen to seek a drink. Unless there is some direct link with the sides any amphibians which are ready to leave will not be able to make their escape. It is a good idea to make one or more access areas down to the pond, so that animals will be able to get down to drink and froglets will be able to get out of the water.

Fountains and waterfalls

Adventurous souls might want to improve their simple pool with fountains and waterfalls. Unlike the 'natural' ponds, those in gardens are generally more shallow. Although the lower regions do not heat up as quickly as those near the surface,

nevertheless, continuous high atmospheric temperatures do cause the water to warm up. With a steady flow from a waterfall or fountain, not only will the water be oxygenated, but the temperature will usually be kept down. This is advantageous to the animals living in the pond, but too great a current is not beneficial to some aquatic plants, such as water lilies, which prefer a static situation, and it is worth remembering that although a fierce waterfall will perhaps put more oxygen into the water it may disturb the plants as well. Trickling water is, if anything, more 'musical' than the continuous splash of a roaring torrent!

In planning waterfalls and fountains the first item of equipment is a pump, but *please remember that water and electricity combine to produce what can be a potentially very dangerous situation.* With this in mind, and for safety's sake, it is vital that a qualified electrician be called in to carry out the work. Better *not* to have a pump than one which would prove dangerous. Having made that point clear, the type of pump employed has to be decided on. Obviously those out of sight give rise to the neatest installation, although it is possible to hide one above the surface. The submersible pump which will be concealed under the water, must have its waterproof lead *sealed* inside it. Since it isn't possible to have satisfactory joints under the water, the lead should be long enough to reach outside the pool. The end of this cable will have to be attached to a weather-proof extension lead to carry the supply from the nearest socket. The connectors used to join this lead to the pump lead should also be of the weatherproof kind. If the pump is to be used for a fountain the job is relatively simple. Once installed, the outlet nozzle must be above the water level and to this must be added a suitable fountain head. The fountain is now ready for use – once you have turned on the electricity!

Surface pumps are generally adopted for use when there is to be more than a single outlet. The advantage of such a pump is that a qualified electrician is not needed for the installation, as long as a competent person checks it. The pump should be housed in a weatherproof container, although it must be such that air can get in. Because the water source will still be the pool you need two polythene tubes. One of these will be from the pool to the pump: the other from the pump to the waterfall. Clips should be used on the tubes to control the flow of water.

It is obviously a good idea to plan the pool and its fountains or waterfall at the same time, but there is no reason why this latter addition cannot be made at a later date when funds allow.

Alternative garden ponds
So far, the pools we have discussed have been sunk into the ground, but in preference to this type you might decide to build a raised pond. There are some advantages in this approach, and one of the most important is that you do not have any excavations to carry out, thus saving quite a lot of physical effort! A raised pond can be made using a preformed fibreglass shape, or with one of the sheets already mentioned, but it is still necessary to prepare the ground onto which the liner or pond shape is going. If a fibreglass unit is being

used, it needs to be supported on bricks which have been covered with polythene to take off their sharp edges. If a liner is being employed, the ground should be covered with two layers of polythene, between which are layers of newspapers. The edge of the fibreglass unit or the polythene can be placed on top of a wall round the edge and held in place with flat-topped slabs.

Of course in the small garden all of these ideas will be out of the question, but if you want to have some water in your garden, you can do so with small receptacles. If you have guttering and water spouts where you can catch the rainwater, you will find that this will give you an area which will be colonized by various sorts of wild animals. Of course it will be too big and too deep for most fish and tadpoles.

One of the favourite ways of making a mini-pond is to use an old sink. If you are able to find a local builder who is demolishing or renovating old cottages, you may be lucky enough to get one of the large rectangular earthenware type which is ideal for various forms of wild pond creatures, including tadpoles. Tin or galvanized baths are also suitable. It is a good idea to treat the outside of the bath with a rust-inhibiting paint. If you are using a sink, you will need to plug the drain hole with clay or cement. Once in position, the sink can be enhanced by putting in plants. Even in the shallow ponds, you can usually manage to encourage some plants to grow, and there are always the floating species, which do not need to be anchored. Rocks will provide shelter for some of your animals. You can stock the ponds as suggested on pages 103-16 for the larger units, although of course the number of plants or animals you have will be much smaller.

Another idea I have seen is an oak cask which has been cut in two, so that it makes two round pools. One or two precautions had to be taken before sinking it into the ground. The metal bands which hold the planks of wood together had to be in the correct position and because the soil, especially when it is very damp, would probably rot the outside of the wood, the cask had to be treated with a non-toxic wood-preservative and the bands with a rust-proof paint, like red oxide. Once sunk into the ground, half barrels could be landscaped, using turf, concrete blocks or paving slabs, but because they are likely to be deeper than a sink, it is a good idea to put water plants in small containers.

Life In The Pond

Having built our pond, the next part of the operation is to stock it. In doing this it is to be hoped that not only will we encourage the normal sorts of invertebrates, which are found in natural ponds, but also aim to encourage amphibians – frogs and toads, and perhaps newts – because their natural habitat is currently vanishing at an alarming rate. By also having shallow areas around the edge of the pond which birds will enjoy for paddling and bathing, we will be doing these creatures a favour, while in the deeper areas of the pool, fish will be able to live and some pond plants will be able to establish themselves.

To make sure that you have a pond in which plants and animals can survive together it is necessary to know something about their ecology. Unlike a field or a wood, the life in a pond is generally more confined: if food runs short or things go wrong, there is no escape for the inhabitants.

The dragon-fly

As in any other habitat, life in the pond forms a delicately balanced community, each species there dependent on the others to a greater or lesser degree. It is only over a period of time that a new pond will become established so that the animals and plants can survive in the environment without the need for man to interfere. Not only is the life inter-dependent, but the very water in which that life lives is also of great importance.

It is a simple, but nevertheless very important, fact that all life on the earth depends on the sun for its well-being. But it is only plants which are able to utilize the energy from the sun to make food. Animal life in the water is hardly likely to flourish unless there are plants. Pure water wouldn't be of any use, and other chemicals which are dissolved in the water are necessary if plants

The water vole

are to flourish. Carbon dioxide is necessary for the plant to make food. Abundant in the atmosphere it readily diffuses into the pond water, making it available to the plants.

The herbivorous animals – the plant eaters – will, in their turn, feed on the plants and thus acquire their share of the energy. Within the structure of the pond there will be carnivorous animals which will feed on the herbivorous species. It is possible that there will also be omnivorous species, which feed on both plants and animals. Then, in addition to the energy derived from the sunlight the green plants will also need organic materials. As the plants and animals die, some will decay: others will be eaten by scavengers. The decaying material releases nutrients essential to the plants when manufacturing their food. It can be seen that the life in a pond is very much dependent on there being a balanced community. This must be in the forefront of one's thoughts and borne in mind when stocking it. Each member of the pond community is dependent on the others. So simple food chains are built up; but there are some animals which feed, not on one kind of animal, but on many, and so the food chain soon becomes a food web, forming a very intricately woven system. If one of the links is removed the whole system could be so easily upset.

From time to time undesirable substances do get into the water, including various dangerous chemicals, and these can bring about the death of many of the inhabitants of a pond. These confined areas of water also tend to suffer more than streams. Where run-off water carries chemicals

The common carp

into a stream the volume of water will normally quickly dilute them, unless the levels continue to be high. The water in the pond is 'stagnant', and poisonous chemicals which flow in build up so that the death of both plants and animals may come about quickly.

Assuming that the necessary gases, as well as dissolved mineral salts, are present in the water, plants will be able to make their food, providing nourishment for the animals which live there. But there is a cycle of events which we must not forget. Neither plants nor animals go on for ever: death comes to all organisms eventually. In a well balanced pond such death is essential. Decay will follow and as it takes place valuable nutrients will be released into the water. But if the pond isn't a well balanced one, then there may be either too much or too little of the nutrients.

In the 'natural' pond, like those found in fields and woods, the balance is usually there. This is not always so in the case of the garden ponds, and, especially where new ones are planned and built, it is necessary to keep a careful eye on the situation, to check that animals are not dying and plants continuing to grow unchecked, otherwise imbalance will soon become a feature.

How To Choose Water Plants

Before you contemplate the purchase of plants you will need to decide which method you are going to adopt as far as the actual planting process is concerned. The bottom and shelves of the pool can be covered with a layer of soil to a depth of about 15 cm (6 in). Plants, including many truly aquatic species, will revel in this unconfined environment, and will grow quickly. Many of the animals you introduce will also flourish in a pond with a naturally muddy bottom, and if you want a conservation-orientated garden you will undoubtedly adopt this method. The main disadvantage is that your pond will be cloudy and muddy for most of the time and the only alternative is to plant in pots and containers, which obviously makes for a tidier pool.

The plants which you select for your pond should encompass some from each of three groups: these are the oxygenators, the free-floating species and the marginal plants, many of which are listed in the table on page 106.

The oxygenators

Although it is useful to have some examples from each of these groups in one's pond, the most important plants of all are the oxygenators. All of these are submerged species, and some authorities simply term them 'water weeds', although their value in providing oxygen cannot be dismissed lightly. There are several different species, but it is as well to check with a supplier as to the best buys for your pool. Some species do well in some situations and not in others, and yet no one seems to be able to put their finger on the precise requirements of each. If you cannot obtain such information or prefer to do your own research you could select one of each species available if the size of your pool allows. You will soon discover which species seem to cope best in your particular environment. It is best to remove the strugglers, and perhaps to some extent those which are particularly prolific, since you might have problems with their unchecked growth.

It is best to plant the oxygenators first. Some examples, including the water crowfoot, tend to spread quickly and in the interests of tidiness, as well as of constricting growth so that the pond does not become overcrowded, it is well worth using containers. Plant pots, crates, or plastic containers are all suitable. If you use crates put sacking in first. As one would expect this will eventually rot but by this time the plants will have grown, and the roots should have formed the soil into a solid block. You will have noticed this in ordinary garden pots, especially when the plant becomes root-bound. Once ready and planted the containers are placed in the bottom of the pool.

Although the plants are restricted by their containers, and the roots won't spread, the foliage may increase too quickly and clog up the rest of the pool. In autumn enough should be removed to keep the pond stable. It is best to remove some roots as well: if you just break off the stems these will grow up again in the following spring.

The floating species
The second group of plants for the pond are the floating species. Their only claim to fame is that none of them needs to root into soil in order to flourish. They have their advantages in that, because they cover the surface of the water, they prevent unwelcome algae from taking up residence.

Some plants in this group have flowers which float on the surface and perhaps the best known of these is the water lily which is rooted but has floating leaves. The disadvantage of having this species in the small garden pond is that the leaves cover a relatively large surface area. If you do decide to have a lily in your small pond, the best time to plant it is during the spring and early summer – from about April to the end of July. No one will argue that when in flower the water lily, along with other surface flowering species, adds a new dimension to the pond, with its beauty; but apart from aesthetic considerations it has other

useful functions as well. There are many species of water animals which need some form of shelter, and the large leaves of this plant provide it for them. A further advantage of having these species in the pond is that particularly during the summer, they act as a shade for the water below. Algae usually grows profusely during warmer weather. When the lily's leaves stop the sun and make it unable to penetrate into the water this problem is to some extent eliminated.

Plants for the water margin
In natural ponds the area around the edge will usually be moist for much of the year, and here the marginal species will flourish. Some common plants in this category include the marsh marigold and the sweet flag and although you will probably only start with a few, given the right conditions they will grow freely and soon provide a dense sheltered area for certain species of wildlife. The plants found along the edge of any pool will fall into one of three categories. The first group will survive and flourish where the ground is permanently waterlogged, and indeed in soil which has a few centimetres of water on its surface. The second group need water, but they will only survive if they are just above the water line. The third group has to be above submerged soil, and the water is obtained by the roots which penetrate the substrate.

You may find that it is difficult to create these conditions around your pool. An undulating surface around the structure might be the answer, in which case the mounds represent the higher ground, the hollows the pools. We have already suggested that you have shelves in your pool to offer various levels to your plants. If you are keen to have a marshy area you should extend the shelf from the normal width of 20 cm (8 in) or so to around a metre (3 ft 3 in). By having a low retaining wall about one brick high where the pool joins the marshy area, you will prevent the soil from seeping back into the pool. The water for the pond needs to be able to flow over the bricks into the marsh. Ensure that the high areas are at least 15 cm (6 in) above the lowest levels. If you already have a pool in your garden and want to make a marshy area, you will find it very difficult to convert or attach one. However, with a little cunning excavation you can make such an area appear as part of your pond.

SOME PLANTS SUITABLE FOR GARDEN POOLS

The plants listed below are grouped according to the role which they play/where they grow in the pond. Many can be obtained from garden centres/aquatic nurseries. Others can be obtained by post, and addresses will be found on page 120.

OXYGENATING SPECIES

Callitriche platycarpa	Water starwort
C. stagnalis	Water starwort
Ceratophyllum demersum	Water hornwort
C. submersum	Water hornwort
Chara vulgaris	Stonewort
Elodea canadensis	Canadian pondweed
Elodea crispus	Canadian pondweed
Fontinalis antipyretica	Willow moss
Hottonia palustris	Water violet
Myriophyllum spicatum	Water milfoil
M. verticillatum	Water milfoil
Potamogeton crispus	Curled pond weed
Ranunculus aquatilis	Water crowfoot

FREE-FLOATING SPECIES

Azolla caroliniana	Floating fairy moss
A. filiculoides	Floating fern
Eichhornia	Water hyacinth
Hydrocharis morsus-ranae	Frogbit
Lemna gibba	Duckweed
L. minor	Small duckweed
L. trisulca	Ivy-leaved duckweed
Pistia stratiodes	Water lettuce
Stratiotes aloides	Water soldier
Trapa natans	Water chestnut
Utricularia vulgaris	Bladderwort

ROOTED BUT WITH FLOATING LEAVES

Aponogeton distachyon	Water hawthorn
Hippuris vulgaris	Marestail
Hottonia palustris	Water violet
Nupha lutea	Water lily
Ranunculus aquatilus	Water crowfoot (also an oxygenating species)

PLANTS WHICH GROW AROUND THE MARGINS

Acorus calamus	Sweet flag
A. calamus variegata	Variegated sweet flag
Alisma plantago aquatica	Water plantain
Apium nodiflorum	Fools watercress
Butomus umbellatus	Flowering rush
Calla palustris	Water (bog) arum
Caltha palustris	Marsh marigold or Kingcup
Caltha palustris plena	Double marsh marigold
Carex spp	Sedge – various species
Cyperus longus	Sweet galingale
Equisetum fluviatile	Water horse tail
Eriophorum angustifolium	Cotton grass
Eupatorium cannabinum	Hemp agrimony
Iris laevitaga	Iris
Glyceria spectabilis	Water grass
Iris pseudacorus	Yellow flag
Juncus inflexus	Rush
Lysimachia vulgaris	Yellow loosestrife
Lythrum salicaria	Purple loosestrife
Mentha aquatica	Water mint
Menyanthes trifoliata	Bog bean
Mimulus luteus	Water musk
Phragmites communis	Reed
Polygonum amphibium	Amphibious bistort
Rumex hydrolapathum	Great water dock
Sagittaria saggittifolia	Arrowhead
Schoenoplectus lacustris	Bullrush
Scirpus albescens	Ornamental bulrush
Sparganium erectum	Bur-reed
Typha latifolia	Great reedmace

Obtaining plants

When first stocking your pool with plants you may be tempted to remove established plants from a local natural pond. You *should not*, however, obtain your plants in this way because apart from the fact that it is now an offence to do so, without first seeking the permission of the owner, you could be doing serious damage to the ecological balance of an established pond. On the other hand, you might be able to obtain free plants from a neighbouring pool enthusiast who has a genuine surplus, but if this is not possible in your locality, the best policy is to obtain plants from commercial suppliers; you may decide to buy your plants locally or by mail order, and a list of aquatic nurseries is given on page 120 of this book. There are advantages to be gained from buying plants from reputable nurseries because a good supplier will always supply explicit instructions with his stock, and for best results this advice should always be followed carefully.

There are often many frustrations during this initial stage of setting up and stocking a pond. It will take a while for the plants to become established, and problems may arise during this time. Apart from the larger plants, which you have willingly introduced, others may appear unbidden and this is particularly true of the more lowly algae. In small pools these species can be extremely troublesome, and very hard to eradicate. Their prime nuisance is that they tend to cause the water to cloud up, and it will also often take on an unpleasant green tinge. Once the larger oxygenating plants have become established, the water tends to clear. If you have looked at rivers and streams, you may have seen a slimy weed almost obliterating the other plants. Known as blanket-weed, this is usually a sign of pollution. It does appear from time to time in the garden pool and while it does not unduly affect the life there when it merely covers the surface of the water, after a short while it tends to 'strangle' the other species, preventing them from functioning efficiently.

Fish For Your Pond

In most garden ponds the choice of fish is often limited to fancy varieties, like goldfish and other hardy cold water species, to the exclusion of natural varieties. If your pond is to be truly natural you should have sticklebacks, minnows, tench and rudd. Obviously you will want to make sure that there is some balance in your pond, and tench are particularly useful fish to include, because they are refuse collectors and eaters. They feed on the bottom and will take care of most, if not all, of the debris. With such a useful species there are bound to be disadvantages. The main one is that tench (and to some extent rudd and sticklebacks) will often outgrow the size of the pond, and get too big for their well-being and at the same time upset the other fish.

Perhaps it would be as well to mention here that one cannot put indefinite numbers of fish into a pond. You will need to work out roughly how much water your pond holds and having determined this factor, you can find out how many fish your pond will take using the table on page 108. Most aquarists will give you valuable information about ponds and stocking them. The generally accepted rule is that one should have 12.5 mm ($\frac{1}{2}$ in) of fish for every gallon of water. So the capacity of your pond determines the size and numbers of fish which you can keep. With a pond holding 20 gallons (90 litres) of water you would be able to have fish measuring 25 cm (10 in) in length. This is where you can benefit by selecting species of fish which will not grow to any great length. Of course, the choice will ultimately be yours. You may find that you prefer rudd and tench to the smaller stickleback. In this case you could plan to move some of them as they grow too big for the pond. You should be able to obtain fry from various sources. Garden and water centres (see the Yellow Pages of your local telephone directory) are often useful sources. Alternatively there are suppliers by post, and some are given at the end of this section. You can also consult specialist publications, like *Aquarist and Pondkeeper* (address on page 120).

Don't be tempted to stock it with more fish than it is capable of holding, because the outcome will inevitably be disappointment. Oxygen will be used up by the fish, and although it is replaced partly by your water plants, and partly because some will diffuse into the water from the atmosphere, if there are too many fish then they will use up more oxygen than there is available. While many of the smaller invertebrates can survive in a situation where there is less oxygen, this is not so with fish. To prevent failures it is important to ensure that the pond is a balanced one, as we have earlier suggested.

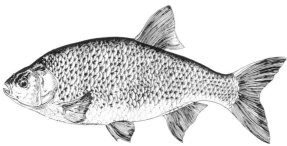

The rudd

The stickleback

The life-cycle of the stickleback is given here in detail because not only is it perhaps the most popular species of fish to be found in the natural pond but its breeding habits form a particularly interesting subject for wildlife observation in your garden. When you were young, and if you lived in the country, one of your delights would probably be going out with a net tied to a bamboo cane, which you used for fishing. And the most likely fish that many people caught, and become acquainted with, was the tiddler, better known in scientific circles as the stickleback.

There are three different species – the three-spined, the ten-spined and the fifteen-spined (named from the spines on the dorsal surface). Even though these spines are meant to offer the fish some form of protection from would-be enemies, they still fall prey to a number of different species. They are one of the favourite foods of the kingfisher, and the otter is also partial to them. The species most often caught will have been the three-spined stickleback, since the ten-spined is rather rare, and the fifteen-spined is found only around the coast of the European continent.

Having acquired some sticklebacks, you will need to make sure that your pond will provide them with a supply of food. As we have already mentioned, the smaller animals are important in the pond, and you should have already stocked your pond with these. Sticklebacks feed on worms and snails, as well as on smaller crustaceans, which are found in the natural waters.

The stickleback is probably well-known because of the male's habit of building a nest in the breeding season, and then guarding it and the young. His other distinguishing feature is the fact that he develops a red patch on his belly. This acts not only as a come-on to the female, but also as a warning to other males to stay away.

At the approach of the breeding season the male

APPROXIMATE GUIDE TO THE NUMBER OF FISH YOUR POND WILL HOLD

Key: * Small fish 5-7 cm (2-3 in)

† Large fish 12-15 cm (5-6 in)

VOLUME OF WATER		APPROX. AREA		NUMBER OF FISH	
LITRES	GALLONS	M²	FT²	*SMALL	†LARGE
455	100	0.9-1.1	10-12	6	2
570-680	125-150	1.1-1.4	12-15	8	3
680-1140	150-250	1.4-2.3	15-25	12	4
1370-1820	300-400	2.3-3.3	25-35	18	6
1820-3200	400-700	3.3-4.6	35-50	24	8
3200 +	700 +	4.6 +	50 +	30 +	10 +

takes on his seasonal colours. Not only does his belly go red, but the back takes on a greenish hue, and the eyes become distinctly blue. In full coat he now selects the area which he will make his for the breeding season. The territory established, generally on the bottom of the pond, he will start his courtship display, which will hopefully attract not just one, but usually several, mates. His territory is his alone, and will be guarded jealously and viciously if intruders approach. Although aggressive when patrolling his patch the stickleback is quite docile when he leaves it. Although he will generally ward off all trespassers he is usually particularly hostile towards the other males.

After the initial stages of establishing his territorial rights, the stickleback has to prepare the nest for the forthcoming breeding season. He is very adept at this. Having first used his mouth to make a dip in the sand he will then place algae in it, and using a tacky substance which is made in the fish's kidneys, he ensures that the plant will be held in place. The main structure completed, there is still one very important task to perform. The male now pushes his way through the tangled mass, until he has a tunnel which goes in one side, and out the other. Now he has to persuade a female to visit his abode and lay her eggs. Although the sole occupant of his territory he is not the only male stickleback in the vicinity looking for a mate. The females are drab when compared to the male in his breeding scales. Exhibiting just a silvery tinge to their scales, and with a belly swollen with eggs waiting to be laid, one or more may swim into the territory. Although his bright colour will attract the female, the male will still come out from his nest and perform his own dance. The females not ready to lay their eggs will show no interest, but those which are will trail behind the male as he makes his way back to his nest. The female takes his invitation seriously, and goes into the nest after him. His aim achieved, the stickleback starts to attack the female gently, which encourages her to lay her eggs. This completed, she leaves so that he can go in and fertilize them. Her task completed she takes no further interest in eggs or male. In fact, as if to make sure that she doesn't bother him, he goes after her, and sees her off his territory.

He is now in sole charge of the future offspring.

His first task is to make any necessary repairs to the nest, which almost certainly became damaged during the egg-laying operation. He checks that the eggs are out of sight so that they will not be attacked by would-be predators. He will act as mother and father: in fact sole guardian of his yet unborn youngsters. He is on regular duty to guard against any intruders, making every effort to drive away even the fiercest of them. The eggs need a constant supply of oxygen, and so, with deft actions, the male makes sure that a current of water passes through the nest.

The male's task of guarding the eggs may last for as long as twelve days or so, or perhaps be over in less than half that time. The deciding factor is the temperature of the water. Once the eggs have hatched the fry will stay in the nest for a couple of

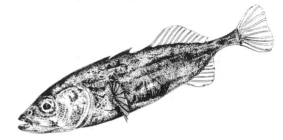

The three-spined stickleback

days. Once out of the eggs the small fish are capable of creating their own current, and so the male gives up this task. It is fortunate that he can, because he now has a family of very wayward sticklebacks to control. This activity will take virtually the whole of his time for about the next fourteen days. As each member of the family escapes from the shoal, the male picks it up in his mouth and carries it back to join the rest of the fish. Growth continues apace, and soon the tiny fish are much increased in size. The protective attitude of the male changes, and after some two weeks he will no longer keep tabs on his straying offspring. His task now over, and the breeding season passed, his once brilliant belly fades to take on the normal colour. He leaves the shoal of youngsters to their own devices and seeks out his own kind. He is unlikely to see the next breeding season, because he will probably fall prey to one of the predators in the pond or stream. In the garden pond he might survive longer than in the natural field pond.

The Smaller Inhabitants
Of The Pond

Many people who establish a pond in their garden are surprised at the life which eventually arrives, as it were, 'out of the blue'. Other people never look closely enough and so are not aware of its existence. But within a short time of the water being put into the pool, the first of the uninvited guests will arrive. These uninvited guests are members of the insect order and there are various ways in which they enter the new environment. One of the first tasks after finishing your pond will have been the introduction of plants. The roots of these plants could have been the source of your newly acquired creatures. Eggs, as well as the animals themselves, cling to the roots of aquatic species. If your pond is large enough, then it may have been visited by ducks, or even, if you are lucky or unfortunate, depending on which way you look at it, by a heron. These creatures often pick up the eggs of aquatic species on their feet and carry them from one habitat to the next.

There are some species which do not rely on other forms of transport, but make their own way to the new pond under their own steam. Some of the species which are usually found in ponds, like the water beetle, the dragonfly and the mosquito, have aquatic stages to their life history, while the adults are able to fly. These will lay their eggs, either in the water, or close to it. Furthermore, there are other species the adults of which will take off in search of new areas of water. This might happen when their own pond is about to dry up. In this way the adults make their way to, and colonize, the newly completed garden pond.

The pondskater

The invertebrates – we use this term because they might not all be insects – can be divided according to the particular part of the pond which they inhabit. We might as well start at the top and work downwards. If you have watched a pond for any length of time you will have seen various insects walking about on the surface of the water – yes, literally walking about. There is a 'skin' on the surface of the water of all ponds, and this can support the weight of small animals. The pond skater has well spread-out legs so that its weight is distributed over a larger area. Pond skaters are quite common in almost all areas of still water, and so ponds are favourite haunts. Pause for a moment and watch these fascinating creatures as they skate or skim across the surface of the water. About 8 mm ($\frac{1}{3}$ in) in length, the pond skater has a narrow body. Using the spread-eagled middle legs they manage to propel themselves across the water. The pair of legs at the front are rather short, and these are the ones which the skater will use to hold its food. The back pair are useful too, as you may be able to observe. These the creature uses for 'steering' its course: in effect they are its rudders. In the normal course of events the skater will skate effortlessly over the water. Try catching one and you will soon realise that, because of their extremely keen eyesight, they are particularly wary of your approaches, darting away, not skating now, but using a leaping action. Normally they appear to move quite well, but once alarmed they are only capable of clumsy movements.

You might expect the underside of the pond skater's body to get wet. This doesn't happen because it has a coating of very small silver-coloured hairs. These trap the air which acts as a barrier against the water. Although living on water, they feed on small animals which live on land. Many such creatures are unfortunate enough to meet their end as they fall onto the surface of the water. Unable to recover and move away, they will provide an adequate source of food for the hungry hunting pond skater. Agile and ever watchful, the animal is useful as a refuse collector, picking up and devouring those animals which would otherwise litter the water.

The whirleygig beetle

At first glance the small black, aptly-named, whirleygig beetle might appear to move about like the pond skater. But there is a difference, although this is not generally apparent at first sight. The whirleygig beetle has part of its body actually in the water and part out of it. The legs are always in the water, although they are seldom seen. Actually you may be able to detect the ripples which spread out from its body as it uses its legs for propulsion whilst swimming. These small legs move about so quickly that they are virtually impossible to see! The beetle's legs are well adapted to its aquatic life; they are flattened and rather broad, for their size. The edges of these swimming legs have hairs on them, which help the animal to 'fix' itself to the water. The legs are like tiny oars, except for the front ones. These are of normal shape, because the beetle uses them for holding on to its food. Although their movements are erratic, they are very much aware. In fact the creature can look above the surface and below it as well. Their eyes are split into two parts: the lower part allowing the beetle to see below the surface, while the rest above the water can see anything in the atmosphere.

The whirleygig beetle seems to travel about the surface of the water in a random, crazy fashion. As they witness the approach of an intruder their reactions become even more erratic! Disturbed from what seem to be their mad movements, the beetles will dive below the surface, taking with them a bubble of air, which contains life-giving oxygen. Your pond will support a few whirleygigs, and you will probably notice how they appear to move about in sociable groups, with each group always avoiding the others.

The mayfly
Some of the most fascinating creatures of the pond are those which have an aquatic stage different in appearance from the adult form. The caddis flies, mayflies and dragonflies all fall into this category. Mayflies are named from the month when most of their numbers emerge. Unlike the other insects the mayfly has an extra moult stage, which makes it unique. The nymphs – the aquatic stage – are quite common in most natural ponds, and if you can transfer them from a friend's established pond to your own you are likely to witness the transformation from an almost insignificant, but nevertheless interesting, nymph to adult form. There are several species of mayfly larvae to be found in the ponds of our island. The way in which they move about in their aquatic habitat varies from species to species. There are those which will burrow their way into the soft muddy banks, so common in many natural ponds, but often lacking in the garden type. Others are not so energetic as they crawl, almost lethargically, over some of the water plants. There are some species which do not live in ponds, but prefer the more demanding conditions in streams. There are others in ponds which are very energetic, and seem to swim perpetually.

You will probably find nymphs of various species once your pond becomes established. Mayfly nymphs have three 'prongs' at the tail end of the body. Generally the most common species are the ones which have a pale greyish-green body colour in the early stages of their life. Once they are fully grown they take on a darker colour, usually leading to brown. Fully grown nymphs will usually measure about 8 mm ($\frac{1}{3}$ in) in length, and are torpedo-like in outline, a shape which

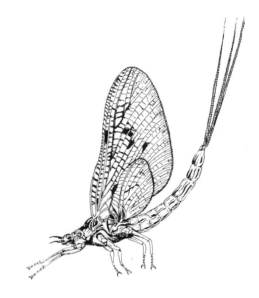

The mayfly (enlarged)

allows them to move with great rapidity through the water. One of the fascinating aspects of their movement is that they appear to start off at full speed, and stop just as suddenly.

Once you have mayfly nymphs in your pond, you will be wise to keep a close eye on the water, particularly during the month of May, so that you will be able to observe them as the adults emerge. Mayflies may continue to leave the pond throughout June, July and sometimes into August. Even though you will probably miss the actual emergence and transformation from nymph to adult in the pond, you could remove some from the outdoor habitat, and place them in a smaller aquarium inside the house. Clear plastic lunch boxes are suitable as they will give clear vision all the way round. One change which takes place is that a layer of air collects under the skin of the insect. This is important because it makes it buoyant so that it will rise from the mass of water to the surface, where it remains without moving. Then suddenly, with a quick twist of the abdomen, the skin splits, and as the mayfly emerges it flies away immediately. This is vital, otherwise it might drown. But unlike the dragonfly, the metamorphosis of which is complete at this stage, the mayfly has still one more stage to undergo. Until its final moult its movements will be restricted, and it is not likely to go far. If it has emerged from an aquarium in the house, it will perhaps fly from the container to alight on a window. In the natural state it will fly from the pond and settle on a suitable resting post nearby, perhaps a branch or tree trunk, out of the water, or maybe on some emergent water plant.

There are distinct differences between male and female mayflies. In those we have described here the females have a variety of shades from a greenish yellow tinge through to a light brown shade. Males are more distinct in colour, being dark brown. Male mayflies have an elaborate courtship ritual, which takes place in a group. From time to time females looking for mates will approach the males. Immediately one enters the cavorting group, it will be grasped by the front legs of a male insect. The male and his mate leave the swarm and fly away. All this happens in the evening from about half an hour before the sun sets. Those males which have not found mates will find a suitable place to rest, and will perhaps repeat their performance the following evening. It is likely that many will not make the following day but, their store of energy expended, will die. Those which have paired will mate, a process which is over in a short while. The male has now used up what little energy his body had, and he drifts to the ground, his life ended. The female cannot give up so easily: she has another very important task to complete. Looking for suitable water in which to lay her eggs, she will take off. A convenient site located, she will dive, and close to the surface an egg will be catapulted into the water. This task continues until all the eggs are disposed of. Her short life's work over, her role fulfilled, she too will die from exhaustion. Most female mayflies will fall onto the surface of the water, and provide suitable food for many species of aquatic life.

Encouraging The Frog

Over about the last ten years or so, the numbers of amphibians and reptiles, such as toads, newts and lizards, have decreased alarmingly. The disappearance of these once familiar and common countryside creatures is due to a number of factors. Years ago when almost every farm had livestock, several fields in each farm complex would have had ponds. Further, one of the typical and expected features of many villages was the village pond. Drainage in and around fields was accomplished by dykes. Alas, water for livestock is no longer necessary for two reasons. First, the number of farms rearing animals has decreased; second, any water still needed is usually piped. So ponds have been filled in, and ditches are no longer necessary because water is piped from the fields.

The widespread use of chemical sprays has led to the deaths of untold numbers of amphibians. Water seeping into the ponds and ditches has resulted in great concentrations of very toxic materials, which have killed both frogs and toads.

With increasing urbanization more and more traffic has led to the deaths of thousands of frogs on roads. Frogs have a habit of returning to the same breeding grounds year after year. Where the concrete skyscraper has replaced open fields, frogs returning to their normal haunts find the water gone.

The habits of the frog

So, how can we help to restore the balance in our gardens? Many garden ponds have been colonized by frogs. Pairs have managed to find their way into these new concrete receptacles, which have proved invaluable as a new habitat for many thousands of frogs. Indeed close to my house, a friend has had no less than a dozen frogs in his garden pond at different times. In another area, my in-laws have had frogs 'taking over' their garden pond, which was recently built for fish. In some cases it is possible that the houses occupy land which was once farmed, and so probably there was a pond in one of the fields. In a desperate effort to find water, without which they will be unable to spawn and perpetuate their own kind, the frogs have come across ponds, probably by mistake, but have seized the opportunity to breed

The common frog

male has found his mate and mating has taken place, the familiar spawn will appear. Frogs spawn appears as a mass: that of toads in a string, so there is no mistaking the two.

With many frogs arriving at and using the same pond you will soon be aware of it should the pond suffer from chemical pollution – which happens if run-off water from the surrounding fields drains into the pond carrying the poisons with it. *All* the amphibians are likely to suffer. If you want to help then contact your local trust for nature conservation (details of each County Trust from SPNC, address on page 7 – please send self-addressed envelope).

How to foster frogs
It is never advisable to take spawn from a pond, but there are a number of firms and organizations which sell spawn. While you might think this is unnatural, if you are conservation-conscious then it might be a good idea to purchase some, so that you can release frogs into your area. There is one word of warning. In certain parts of the country you will never find spawn, or get it to develop, because there is a lack of iodine in the water, something which the developing tadpoles need.

If you don't have a pond in your garden, and would probably not normally have the pleasure of having frogs spawn in your garden, don't despair, because you can help to increase the frogs by obtaining spawn. To rear it, all you will need is some form of waterproof container. You might have a sink or a bath which you could make into an outdoor pond, and these will prove ideal. Failing this, a large bowl indoors will serve the same purpose. Actually there are advantages in using this because you will be able to follow the fascinating change from spawn to froglet.

You will not need very much spawn. In a mere handful, or less, you will have a plentiful supply of eggs. You will also need to take some water from a local pond – preferably one which already has some spawn in it, as this is unlikely to be polluted. Purchase some water weed from a local garden centre or water garden. This is necessary for the tadpoles, when they hatch. Once you have your spawn you can tip it into your garden pond, your bowl, or your sink. If it is going into the garden, don't forget that fish and frogs are generally not compatible.

there. Other frogs seem to have discovered them when looking for winter hibernatory quarters. But why other areas have been chosen is not really known, and is rather mystifying. It may be that frogs have been hibernating in other parts of the garden unbeknown to the owner for many years.

Although there might be frogs in your pond, the chances are that they will not breed there. The ways of wild creatures are devious and not easy for us to understand. Although the whys and wherefores are not known, naturalists are well aware of the 'spring frog phenomenon'. Given the correct conditions at this time of year the frogs will leave their winter sleeping quarters and make their journey to their breeding ponds. How far they travel, where they come from and why they all seem to arrive at about the same time are questions to which we do not have any reliable answers. It is known, for example, that should the frogs arrive 'early' when a further cold spell is anticipated, they will postpone their mating activities until conditions are likely to be more favourable.

The frogs' mating ponds are seldom kept secret for very long. Once these amphibians have arrived their monotonous, and familiar croaking chorus, made by the males to attract their mates, will soon be heard. Alas, this sound, once so common in the countryside, becomes rarer every year. The strange thing about the frogs is that while some ponds may support large numbers of these mating amphibians, other ponds, similar to them to all intents and purposes, do not have a single specimen! Once the

From spawn to tadpole and froglet

Even if you have a pond, once you have bought or collected your spawn, you will probably want to keep some inside so that you can actually see the tadpoles developing. If you are keeping the spawn in a container, whether inside or out, it is a good idea to cover the top with a net. This will prevent marauders, and in particular cats, from getting at it. Your small pond, whether inside or out, should not be in direct sunlight.

Unless you have actually seen the frogs mating and the eggs being laid you will not be sure of the stage of development when you either buy your spawn, or discover it in your own pond. However, with the help of a magnifying glass you will be able to look more closely at the eggs. If we assume that you obtained your eggs soon after they were laid, within a few days the black 'blob' inside the egg will take on a comma-shape. It must be remembered that although the eggs were all laid at the same time they are likely to develop at different rates, and you will also probably find that there are some infertile eggs which do not develop at all.

Other features will be seen as gradually a tadpole begins to take shape, and movement will become apparent as the developing creature starts to wriggle inside the jelly. Within ten days or so after the eggs were first laid – temperature does affect the time taken for development – the tadpoles will have made their way out of the egg and through the jelly. At first they appear inactive: they might find a piece of water weed to cling to, or hang on to the remaining mass of spawn. Although they may wriggle about they do not feed for the first few days, because they do not have any mouths. The egg sac will remain attached to them and this provides them with their necessary nourishment. Over the next weeks you can watch the tadpole as it makes the transition from its aquatic form to a fully fledged froglet.

The remaining food in the yolk will not nourish the frog for very long and the mouth will start to function after a day or two. Their eyes will open as well. No longer do they need to cling to one weed, for now they are able to swim about and are capable of an independent life. You will probably find that all your tadpoles make for the water plants. A fallacy is that the tadpoles are actually feeding on the plants: they are actually eating the

The life cycle of the common toad. The adult is shown, together with the eggs and the developing tadpoles

algae, very small plants, which are attached to the leaves of the larger plants. If you happen to have a microscope you will be able to see this often virtually covering the complete leaf surface. At this stage, the tadpole is fully aquatic. It can only survive if it lives in water. Here, with the aid of its feathery gills it is able to extract oxygen from the water. Slowly the external gills vanish and the internal ones will take over their important work. These gills are rather like those of fish, and it is possible to see small slits in the side of the head. The rate of development is affected by temperature. In natural conditions in a pond when the temperature is down, each stage through which the tadpole passes before becoming a froglet will be prolonged: warmer weather tends to speed up the process.

The animal will use its horny jaws to scrape the surface of plants as it continues its feeding activities. For the first six weeks or so, the creature looks like the tadpole which we know so well. Then changes take place, and these are quite rapid. Where the tail joins the body rear legs will appear. You can watch them grow: at first they start off as what can best be described as buds. These soon sprout until they resemble legs. The front legs will start to appear. During this period of quite rapid growth the tadpoles will shed, or slough, the skin.

Their feeding habits have also changed drastically. Now they will be carnivorous, although it is difficult to say when this actually happens, and it varies, since the tadpoles develop at different stages. As carnivores they will need a supply of freshly chopped meat. Soon after the front legs appear it is worth dropping some in. Small amounts at a time are important, and any uneaten material should be removed because it quickly decays, and pollutes the water. Provided that there are small water animals in your garden pond the tadpoles will have a supply of natural food.

Once the front legs have appeared other changes will take place. The most noticeable of these is the gradual shrinking of the tail. While these external changes can be seen, changes are taking place internally too. From about eight weeks the gills start to disappear, and the tadpoles, now beginning to appear frog-like, will be developing lungs. You will be able to watch the disappearance of the gills, although you can't actually see the internal lungs, but the fact that a change has taken place will be apparent because the tadpoles will come to the surface from time to time to take a gulp of air. The tadpole shape becomes less obvious almost daily, and the frog form more so, and quite soon a miniature frog will appear in your 'aquarium'. Now the animal needs a platform ready to make its getaway – as it would do in the wild. You can put in a piece of stone, a brick, or even a sod of grass. Once on this the froglet will want to use its powerful legs to leap to freedom, so ensure that the top of the container is covered now, if it wasn't previously.

This stage completed you have done all you can for your amphibians. Now they need freedom. If you have a garden pond you can put them there, as long as you bear in mind the fish risk. Alternatively, you can take them back to a local pond. Whether the amphibians which you introduce into your pond will actually use it for breeding purposes remains to be seen. If they don't you can obviously continue your conservation-oriented activities in subsequent years.

The froglet becomes a dry-land creature
Once the froglets are about three months old, they are ready to leave the aquatic habitat for a terrestrial existence. They will wait until there has been a shower of rain before they actually make the transition from water to land. In a pond which has been sought out by many frogs for their breeding activities, the large numbers of froglets which leave at one time after rain has given rise to the belief that they have fallen with the raindrops. As with all wild creatures, the percentage which reach this stage, let alone make it to adulthood and maturity after three years, has always been quite small. Birds and other animals, as well as various reptiles, will soon make short work of the tiny froglets, and the population will quickly decrease. This is in addition to those which have already died during the tadpole stage as a result of attacks by natural predators.

The frog's diet
In their natural state the survivors will find plenty of food in the form of a wide variety of insects. The frog has a long, sticky tongue, and unlike most animals this is not attached to the throat, but to the front of the mouth. As it is flicked out, it will trap insects, and the tongue is then folded

back so that the insect is directed towards the animal's throat. Its prey is swallowed whole. As well as insects and worms, slugs also feature in its diet. It can be seen how valuable an asset frogs are to the gardener, as well as to the farmer, taking as they do vast numbers of harmful invertebrates and keeping their numbers under control.

If you have a garden pool, you will want to know something about the ideal conditions if your tadpole-rearing, frog-breeding activities are to have a chance of success. Frogs need to spawn in shallow areas, whereas toads prefer deeper water. A depth no greater than 60 cm (2 ft) is suitable for toads and under 30 cm (1 ft) for frogs. Plants are important in the scheme of things. Initially, in the case of toads, because the eggs are strung around these, and later because the tadpole will need them for food.

If you are going to encourage the tadpoles and hope for their survival, then, unfortunately as far as some people are concerned, fish cannot be kept in large numbers in the confined space of a garden pond – indeed, perhaps it is best to exclude them altogether. If they are left in, whilst they will probably be unable to attack the spawn, because of its slippery, protective jelly, they will certainly be able to attack the tadpoles and will do so with relish.

In fact there are many naturalists who are altogether opposed to the idea of stocking fish in the garden pond on the grounds that they would eat all the insects and other invertebrates.

No one actually knows how a frog 'decides' that a site is suitable for breeding purposes. Indeed their choice is sometimes inappropriate, because the site may dry up in summer. Often quite close at hand there is a much better pond, which the frogs have spurned when choosing their breeding site! In dry summers when some of their breeding sites dry out, the tadpoles not yet ready to leave the water will die.

Garden ponds are generally more suitable for frogs than toads because their size makes them unable to support large numbers. In some areas toads in the wild have been seen in breeding colonies containing over one thousand individuals. Since frogs breed in smaller numbers the garden pond can help to save them from extinction, a fate which may be more imminent than we think, though twenty years ago it was inconceivable.

Amphibians And Reptiles Which May Be Found In The British Garden

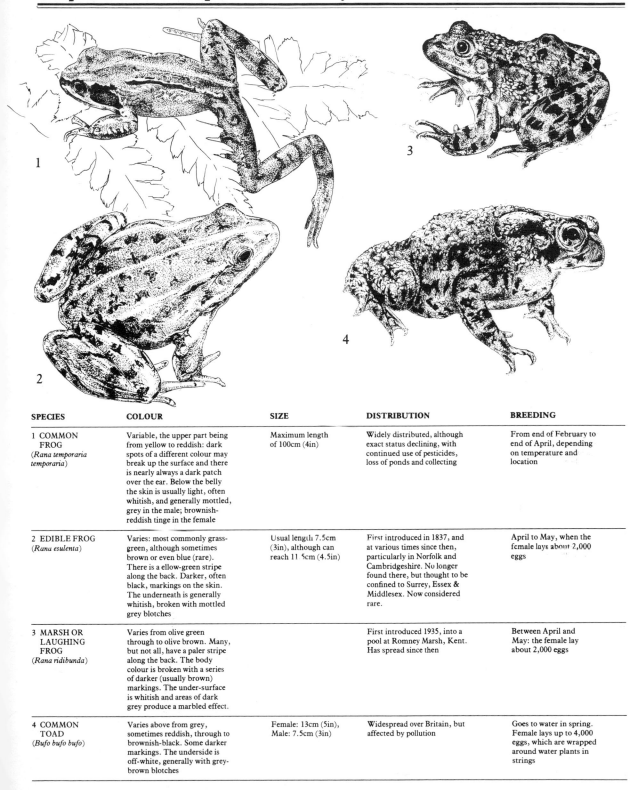

SPECIES	COLOUR	SIZE	DISTRIBUTION	BREEDING
1 COMMON FROG (*Rana temporaria temporaria*)	Variable, the upper part being from yellow to reddish: dark spots of a different colour may break up the surface and there is nearly always a dark patch over the ear. Below the belly the skin is usually light, often whitish, and generally mottled, grey in the male; brownish-reddish tinge in the female	Maximum length of 100cm (4in)	Widely distributed, although exact status declining, with continued use of pesticides, loss of ponds and collecting	From end of February to end of April, depending on temperature and location
2 EDIBLE FROG (*Rana esulenta*)	Varies: most commonly grass-green, although sometimes brown or even blue (rare). There is a ellow-green stripe along the back. Darker, often black, markings on the skin. The underneath is generally whitish, broken with mottled grey blotches	Usual length 7.5cm (3in), although can reach 11 5cm (4.5in)	First introduced in 1837, and at various times since then, particularly in Norfolk and Cambridgeshire. No longer found there, but thought to be confined to Surrey, Essex & Middlesex. Now considered rare.	April to May, when the female lays about 2,000 eggs
3 MARSH OR LAUGHING FROG (*Rana ridibunda*)	Varies from olive green through to olive brown. Many, but not all, have a paler stripe along the back. The body colour is broken with a series of darker (usually brown) markings. The under-surface is whitish and areas of dark grey produce a marbled effect.		First introduced 1935, into a pool at Romney Marsh, Kent. Has spread since then	Between April and May: the female lay about 2,000 eggs
4 COMMON TOAD (*Bufo bufo bufo*)	Varies above from grey, sometimes reddish, through to brownish-black. Some darker markings. The underside is off-white, generally with grey-brown blotches	Female: 13cm (5in), Male: 7.5cm (3in)	Widespread over Britain, but affected by pollution	Goes to water in spring. Female lays up to 4,000 eggs, which are wrapped around water plants in strings

SPECIES	COLOUR	SIZE	DISTRIBUTION	BREEDING
5 NATTERJACK (*Bufo calamita*)	The upper surface is olive-grey to olive-green, with variable markings from greyish to reddish. The underside is off-white, with dark grey blotches. The female takes on bluish tinge to throat in breeding season.	Female: 10cm (4in), Male: 8cm (3in)	Now very rare	From April through to September in water. Female lays up to 4,000 egg
6 GREAT CRESTED NEWT (*Triturus cristatus cristatus*)	Brownish-black above, usually with darker spots. Small white spots on flanks. The underside is yellow to orangish, broken by black blotches	Female: 15cm (6in), Male: 13.5cm (5.5in)	Widespread over the British Isles	Mating takes place in water from March to July. Up to 300 eggs are laid, which hatch in 2-3 weeks
7 PALMATE NEWT (*Triturus helveticus*)	Olive-green to olive-brown, usually broken by small spots. Female has a reddish stripe down the back when on land. Below, the body is yellowish	Female: 9cm (3.5in), Male: 7.5cm (3in)	Widespread, although local in some areas	In water from middle of March through to middle of June. Eggs are laid which hatch after 2-3 weeks
8 SMOOTH OR COMMON NEWT (*Triturus vulgaris vulgaris*)	Upper surface usually brown to olive-brown: there are dark rings on the male and the female has smaller spots. Tail in male has orange tinge in breeding season. The underside varies from yellow to orange, with darker blotches	Female: 9.5cm (3.75in), Male: 11.5cm (4.5in)	Most common of our newts, and widely distributed	As for Great crested

9 THE SMOOTH NEWT (above) and the GREATER CRESTED NEWT, shown in their mating coats

5

9

6, 7, 8

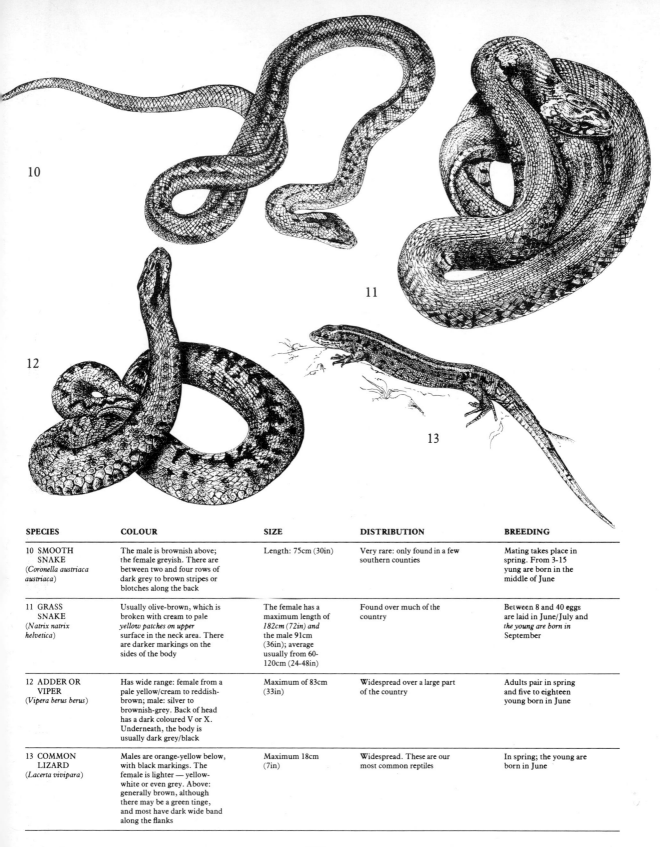

10

11

12

13

SPECIES	COLOUR	SIZE	DISTRIBUTION	BREEDING
10 SMOOTH SNAKE (*Coronella austriaca austriaca*)	The male is brownish above; the female greyish. There are between two and four rows of dark grey to brown stripes or blotches along the back	Length: 75cm (30in)	Very rare: only found in a few southern counties	Mating takes place in spring. From 3-15 yung are born in the middle of June
11 GRASS SNAKE (*Natrix natrix helvetica*)	Usually olive-brown, which is broken with cream to pale *yellow patches on upper* surface in the neck area. There are darker markings on the sides of the body	The female has a maximum length of *182cm (72in) and* the male 91cm (36in); average usually from 60-120cm (24-48in)	Found over much of the country	Between 8 and 40 eggs are laid in June/July and *the young are born in* September
12 ADDER OR VIPER (*Vipera berus berus*)	Has wide range: female from a pale yellow/cream to reddish-brown; male: silver to brownish-grey. Back of head has a dark coloured V or X. Underneath, the body is usually dark grey/black	Maximum of 83cm (33in)	Widespread over a large part of the country	Adults pair in spring and five to eighteen young born in June
13 COMMON LIZARD (*Lacerta vivipara*)	Males are orange-yellow below, with black markings. The female is lighter — yellow-white or even grey. Above: generally brown, although there may be a green tinge, and most have dark wide band along the flanks	Maximum 18cm (7in)	Widespread. These are our most common reptiles	In spring; the young are born in June

Further Useful Information

Organizations and Societies

Freshwater Biological Association, The Ferry House, Far Sawrey, Ambleside, Cumbria, LA22 0LP

Produces guides to various groups of freshwater animals: also sells equipment, including nets.

Save the Village Pond, Bell House, 111-113 Lambeth Road, London SE1

Concerned with the plight of village ponds – also produces wallcharts and literature in connection with the campaign.

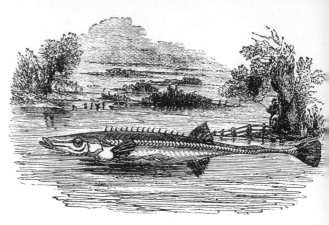

How to Begin the Study of Amphibians and Reptiles, BNA

How to Begin the Study of Pond Life, BNA

Kaye, Reginald, *Modern Water Gardening,* Faber

Mellanby, H., *Animal Life in Fresh Water,* Methuen

Muus, J. and Dahlstrom, P., *Freshwater Fishes of Britain and Europe,* Collins

Nature Conservancy Council, *Ponds and Ditches*

Save the Village Pond Campaign, Book about ponds, address earlier

Smith, K., *Popular British Freshwater Fish,* Jarrold

Smith, M., *The British Amphibians and Reptiles,* Collins

Steward, J. W., *Tailed Amphibians of Europe,* David & Charles

Aquarist and Pondkeeper, Buckley Press, The Butts, Half Acre, Brentford, Middlesex

The magazine is published monthly, and available from newsagents or by subscription from the publishers.

Books, Leaflets, etc.

Bagenal, T. B., *Observer's Book of Freshwater Fish,* Warne

Burton, R., *Ponds: Their Wildlife and Upkeep,* David & Charles

Busche, E. M., *A Handbook of Water Plants,* Warne

Clegg, J., *Freshwater Life of the British Isles,* Warne

Clegg, J., *Observer's Book of Pond Life,* Warne

Dyson, J., *The Pond Book,* Penguin/Kestrel

Hammond, C.O., *Dragonflies of Great Britain and Ireland,* Curwen Books

Equipment and Supplies

Blagdon Water Garden Centre Ltd., Church Street, Blagdon, Avon:
  Liners

Butylmade Ltd, Lloyds Bank Chambers, High Street, Lingfield, Surrey:
  Long life pool liners

Gordon Low Fabrications Ltd, Keppard House, Place Road, Cowes, Isle of Wight:
  Pool liners

Honeysome Aquatic Nurseries, The Row, Sutton, Ely, Cambs, CB6 2PF:
  This firm produces a catalogue of their water

Leisure Laminates, Rawreth Industrial Estate, Rayleigh, Essex:
  Glassfibre garden pools

Maydencroft Aquatic Nurseries, Gosmore, Hitchen, Hitchen,
Herts:
  Water lilies and aquatics

Reginald Kaye Ltd, Waithman Nurseries, Silverdale, Carnforth, Lancs:
  Herbaceous plants and aquatics

Perry's Hardy Plant Nurseries, Enfield, Middlesex:
  Water plants

Robb, S. & Son, St Ives, Huntingdon, PE17 4PB:
  Pool lining materials

Stapeley Water Gardens Ltd, London Road, Stapeley, Nantwich, Cheshire:
  Free handbook about water gardening; sells everything – pumps, liners, etc

Transatlantic Plastics, Garden Estate, Ventnor, Isle of Wight:
  Garden pool lining materials/liners

plants. Furthermore the proprietors are prepared to offer the necessary advice where possible. There is a section devoted to various plants which are needed in a pond.

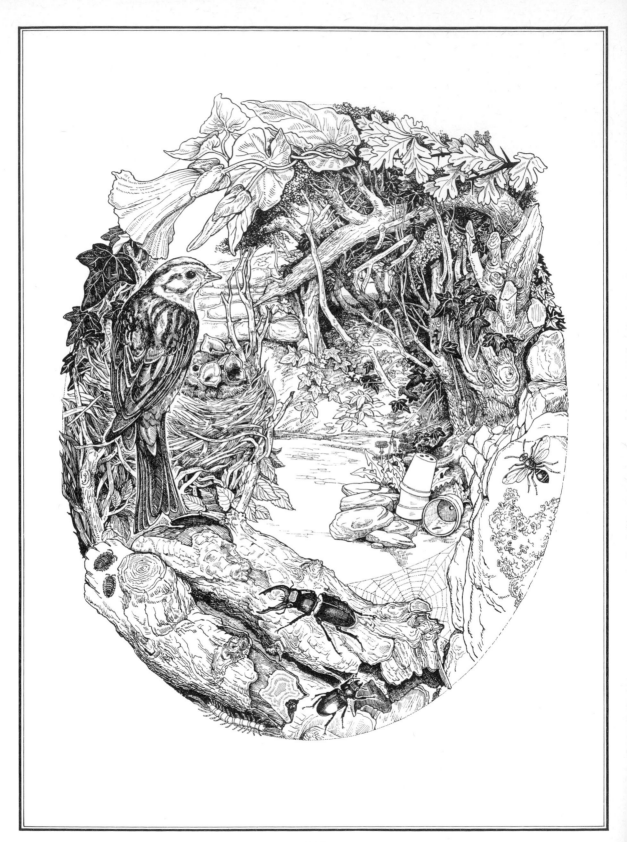

THE
WILDLIFE
GARDEN
MISCELLANY

The illustration shows (from top left to top right):
Convolvulus, yellowhammer nesting with chicks, rotten
wood on which are two woodlice and a centipede, male
and female stag beetles, the web of a garden (or orb)
spider, a solitary wasp (on the stone wall, near lichen), a
section of part of an old hedgerow

Trees In The Garden

A recent report informs us that of all the member countries of the European Economic Community, Britain has the smallest number of trees per square mile. Why has Britain lagged so far behind? Mainly because many of our trees are very old, and very few new ones have been planted. Not only do older trees get attacked by diseases, but they also suffer especially during severe gales. Recently in one small length of hedgerow I know, *thirteen* trees were felled by the wind in one storm alone.

Although many gardens will not be large enough to support the oak which, when fully mature, may measure between 18 and 40 m (60 and 130 ft) in height and have a trunk which is 3 m (10 ft) around, there are smaller species which will enhance your patch, even those you can grow in pots and tubs on your patio (see pages 126-7).

Perhaps it would be wise to look at the value trees have for us and the rest of the creatures which are part of the web of life. Not only do they provide wood: trees are oxygen-givers. As they make their food, they use up carbon dioxide, and give off oxygen, the vital gas which we, and other animals, need in order to stay alive. They are home to a wide variety of not only animals, but plants as well. A tree is a habitat in itself, and although some birds will perhaps only nest there, moving to other places for their food, there are other animals and plants which depend on it, and never leave it. For example, it has been estimated that no fewer than *three hundred species of insects alone* are associated with and dependent upon the oak.

In deciding to plant trees in our gardens there are many factors which have to be taken into consideration. The fastest growing trees are the conifers, but in general the wildlife which they support is rather more limited than that which is associated with hardwood species. When compared with some of our native hardwoods the conifers are much less spectacular as providers for wildlife. Obstacles, both above and below the ground, must not be overlooked. For example, a fully grown oak may reach height of around 30 m (100 ft) and have a spread of from 9 m (30 ft) to 12 m (40 ft) across. You can see that if such trees are positioned incorrectly they could eventually inter-fere with overhead power cables or telephone wires, and the roots will also be extensive. If placed in the line of underground pipes, like those for sewage, they could do untold damage. Imagine what would happen if they were placed close to your house; or the damage which the roots could do to, for example, a neighbour's garden.

Once you have decided on the position and size of your tree or trees, your next decision is on the choice of species. To help you, we give details in table form on pages 126-7, together with some notes about soils etc, to help you decide on a suitable tree.

Some of the larger hardwoods, like ash, oak and sycamore, take a long time to grow to any size. When planning to include these it is also a good idea to include species which are known to grow more quickly. In Britain we have one deciduous tree with needles, which it sheds in autumn. This is the larch, of which there are two species, the Japanese and European varieties. There are several species of wildlife associated with this tree, although its fauna will never be as rich as that of broad-leaved deciduous species. Its usefulness in the garden is that it grows very quickly, and a growth rate of between 50 and 100 cm (15 in and 3 ft) a year is not uncommon.

We have mentioned the conifers, mainly because they are fast growing. Despite the fact that they attract less wildlife than deciduous trees, because of the fact that they are evergreen and so have leaves all the year round they are likely to attract some of the earlier nesting birds.

Planting And Maintaining Trees

Unfortunately many people assume that once a tree is planted that is the end of the matter. But this is not so, and it is worth going to some trouble to make sure conditions are right. One of the most important aspects is obviously the soil. It must be borne in mind that soils in various parts of the country will have differing properties. Careful attention to detail in the first few years of life will get the tree off to a good start. Soils fall into four main groups. Where *loam soils* feature no action need be taken. These soils contain all the necessary ingredients for good growth, having a good mixture of sand, humus and clay. Drainage

124

NUMBER OF INVERTEBRATES ON SOME COMMON TREES SHRUBS AND WILD PLANTS

Naturally, as these trees, shrubs and plants disappear there are fewer food sources for the wildlife population.

BOTANICAL SPECIES	APPROXIMATE NUMBER OF INVERTEBRATES
Alder	about 90
Ash	about 40
Beech	over 60
Birch	225
Bramble	more than 10
Broom	over 10
Buckthorn	more than 15
Crab apple	about 90
Elm	in the region of 80
Fir	about 15
Gorse	in the region of 12
Hawthorn	nearly 150
Hazel	70
Holly	only about 7
Hornbeam	nearly 30
Larch	nearly 20
Lime	just over 30
Oak	over 300
Poplar	about 100
Scots pine	over 50
Spruce	over 35
Willows & sallows	more than 250 (for all species)

PLANTING GUIDE TO TREES FOR YOUR GARDEN

When considering planting trees for your garden the following table, giving average heights, time to maturity, etc., will be useful.

SPECIES	HEIGHT		MATURITY	SOIL	NOTES
	m	ft	yrs		
Alder	28	90	80	Preference for wet soils/grows in most areas	Grows quickly
Ash	28	90	80	All soil types	Grows quickly – surface rooting is a problem
Beech, Common	28	90	150	Mainly chalk soils/tolerant of others	Grows quickly: has deep shade. Colourful in autumn
Birch, Silver	18	60	60	Prefers acid or neutral soils. Likes light sandy loams	Only light shade: does better in groups
Cedar, Atlas	36	120	125	Does best on loams, which are well-drained and neutral. Reasonable on light sands	Evergreen, takes better when planted as a young specimen
Cherry, Wild	15	50	40	Grows on most soils/preference for alkalines	
Chestnut, Horse	28	90	110	Wide variety: prefer deep loams	Heavy shade and deep roots
Chestnut, Sweet	21	70	90	Best on sands – well-drained. Stunted on peats, very chalky and clay soils	Has pleasant autumn tints
Crab apple	9	30	15	Will grow well on most soils	Early flowering
Cypress, Leyland	28	90	45	Will grow well on most soils/tolerates chalk well	Fast growing: can prove useful for hedges. Does well by the sea
Elm, English	28	90	100	Wide range: does better on well-drained, slightly alkaline types	Subject to Dutch Elm Disease
Elm, Wych	30	100	80	Does well on soils with limestone below: needs well-drained situations	Does well near coast, but also subject to Dutch Elm Disease
Fir, Douglas	35	115	100	Many soil types, including acidic, provided that they are moist	Evergreen species: does better when planted young
Hawthorn	9	30	50	Most soil types, except very acid/very wet	A very good hedging plant, although does well on its own
Holly	12	40	70	Grows well on any soil type/very tolerant	Slow-growing evergreen: does well in coastal areas
Hornbeam	18	60	100	Thrives on wide range in lowland areas	Can be used instead of beech
Larch	30	100	95	Does not like dry or poorly drained soils	Grows quickly, but must be planted young
Lime (T. platyphyllo)	30	100	80	Many soil types: thrives best on moist ones	Has a nasty drip – honeydew
Maple, Field	10	36	50	Tolerates any soil type	A hardy tree which is easy to grow
Maple, Norway	21	70	50	Tolerates most soils	Grows well to form a useful screen. Very good autumn tints
Oak, English	18	60	100	Tolerates many soils. Does best on heavy loam or clay	Grows slowly, although it is shrub-like in sandy areas
Oak, Evergreen	28	90	150	Similar to English Oak, but does better on moist soils	Grows slowly. Does well by the coast: good for windbreaks

SPECIES	HEIGHT		MATURITY	SOIL	NOTES
	m	ft	yrs		
Oak, Turkey	36	120	70	Almost anywhere, apart from chalk. Stony loam preferred	Grows quickly, and does well along the coast
Pine, Black	30	100	95	Will do well on almost any well-drained soil. Not very tolerant of damp conditions	An evergreen species, which does better when planted small
Pine, Scots	25	80	70	Does not do well on either badly drained or wet soils. Prefers well-drained acid type	An evergreen species which does well when planted small. Likes coastal conditions
Plane	28	90	90	Prefers neutral types, especially where gravels or loams feature	Useful because it does not have deep shade
Poplar, Lombardy	28	90	70	Gravelly loams favoured, especially where light and exposed	Fast-growing: avoid planting near buildings
Rowan	8	26	30	Well-drained soils preferred, especially where light and composed of light sands or peat	Good production of berries in autumn. Early flowering: does not grow too high
Sycamore	25	80	60	Although will grow in most soils, does well where gravels occur on clays	Is quite quick growing. Does well along the coast
Thuja	30	100	100	Almost neutral sandy loams, which are well-drained are preferred	Does well in wet areas. Grows quickly to provide a good screen
Walnut	29	95	100	Grows best where loams are found on chalk: docs badly on acid soils	Gives good shade
Wellingtonia	30	100	50	Does well on light and heavy soils, except poor chalks	An evergreen which grows well for the first 30/40 years
Whitebeam	12	40	80	Does best in limestone areas	Very good cover, with a dense crown
Willow, Weeping	15	50	25	Prefers areas where there is running water. Does poorly on clay soils	Not really suitable for the small garden
Willow, White	25	80	40	Although does better on light soils, will cope with heavy ones	Grows very quickly
Yew	14	45	100	Cannot tolerate very acid soils, but does well on almost all other types	Although slow-growing is very good for hedges. An evergreen species: plant when young

in this soil type is also very good, but not over-fast. *Clay soils* consist of very small particles of sticky material, which cling together. Besides being difficult to manage, they also have very poor drainage qualities and the surface becomes water-logged. By adding suitable quantities of either sand or peat – or a mixture of both – the lumps are broken up and drainage is improved. Air as well as water is important to the plant, and when the soil is broken up air will be able to get to the tree's roots. In *sandy soils* the particles are large, which means that water drains through very quickly. In very dry conditions, the tree roots will not have enough water. To remedy the situation it is necessary to dig in peat and good farmyard manure. This helps to make sure that some of the water is retained and adds nutrients to the soil. The final type of soil, mainly peculiar to the fenland region of eastern England, consists of large amounts of dead and decaying plant material. This is rich in nutrients, but it is necessary to improve the drainage and this can easily be remedied by the addition of sharp sand.

Seeds or trees?

Having thought about the possibilities for trees and the question of soil, one has to decide whether the trees are to be grown from seeds, seedlings, or established trees. There are of course distinct advantages in selecting a tree which is a metre (3 ft) or so in height, because it provides a visible sign, and, if it is a berried variety, will provide food for animals and birds fairly quickly. On the other hand, it is a delight to watch the slow, but miraculous, transformation from seed to seedling.

Part of the fun of growing the trees from seed is that it is possible to collect the seeds yourself in the autumn, but if you are not able to do this there are a number of seedsmen who will provide good, viable seeds. The easiest to find yourself are those which are relatively large in size: oak, horse chestnut, beech, plane, maple, sycamore, sweet chestnut and yew for example. Whether they will be viable or not is another matter, as the crop of seeds varies from year to year. When collecting seeds it must be remembered that the trees provide a valuable source of food for wild creatures. Taking a few of the seeds is of no consequence but if everyone went and collected vast quantities then the birds and mammals would probably suffer.

You will need to check the seed which you are collecting to make sure that it hasn't already been damaged by animals. In addition, ripe seeds are more likely to germinate than green, unripe ones.

Natural ripening

It is better to let some seeds ripen naturally. With seeds that have hard, protective shells it is best to plant them immediately. You can use pots or trays, filling them with a rich, loamy peat soil. Other, smaller seeds which do not have this covering will remain dormant until the warmer weather arrives the following year. By placing your own hard seeds – the oak, sweet chestnut and horse chestnut – in pots of soil as suggested, you enable the breaking down process to begin, as in the natural state. One of the necessities is to keep the soil moist, avoiding over-watering as this will almost certainly cause the seeds to rot with the result that they will not germinate.

Seeds which do not need this treatment should be dried naturally, and then placed in tins or polythene bags until the following spring. It is as well to give the seeds a good start by placing them in either a branded or home-made seed compost. Germination will be much quicker if the seeds are kept either under glass or in the house. Acorns and chestnuts will need 25 cm (10 in) pots.

The seedlings need placing out in the open so that they can harden off. It is a good idea to leave them in pots outside for about a year. This will bring you round to the first spring. At this stage, the trees can be removed from the pots and placed in a prepared piece of ground. A hole big enough to take the seedling and its soil should be made. You can tip the tree out of the pot, because if it has grown well there will be a good rooting system, which will have bound the soil together. In the garden the seedlings need to be planted about 50 to 60 cm (20 to 24 in) apart in a nursery bed. Although they need to be left here for three to four years so that they will be a suitable size for placing in their ultimate position, it is not a bad idea to dig them up each year, so that the roots can be trimmed. If you don't do this you will almost certainly find it very difficult to dig them up after four years, because of the growth and spread of the roots. If you prefer not to dig up the tree, you can control the root spread by using a spade, but do make sure that it has a sharp blade, otherwise you will damage the roots, and go round the tree to keep the roots in check.

Once your trees are planted out, you can encourage their growth by feeding them in spring. Although the quick-acting artificial fertilizers are tempting to use, give these a miss and go for the natural ones. You can top-dress by using garden compost, if you have some, or peat. If you can obtain *old* manure, this is also suitable.

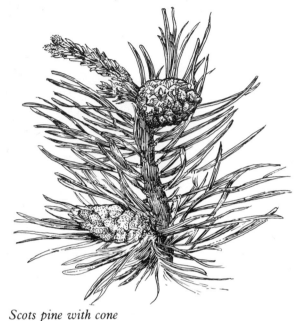

Scots pine with cone

What and how to plant

Deciduous trees have a resting period from autumn through to spring and this is really the only suitable time to plant them, unless you are buying container grown ones, in which case it is possible to plant them all the year round. Apart from the fact that it will generally be difficult to dig a hole when the ground is frozen hard, when the soil is like this planting should be avoided. Conifers are active all year round and the best time to move or plant them is in the spring. Great care should be taken that the soil and the roots are disturbed as little as possible. Again, many of the conifers can be purchased in containers. Although this is more expensive you may decide that the convenience is worth the extra cost.

Preparing the site

Unless there is some reason for purchasing the tree before a site is ready it is best to plant the tree, as opposed to seedlings, in the area where it is going to stay. However, when planting is only temporary a tree should be placed in soft, damp soil. With large trees a hole needs to be dug, and care should be taken to ensure that not only is it wider than the spread of the roots, but also deeper. Once you are certain that you have excavated the soil to the correct depth, you will now need to loosen it for a depth of about another 15 cm (6 in). Artificial fertilizers should not be put into the hole. However, it is a good idea to dig in a small amount of peat, natural compost or bone meal.

Careful support

The stem of the young tree needs a support, which will prevent damage, particularly in gale conditions. A strong stake is needed and this should be driven into the ground for about 30-40 cm (12-15 in), with enough above ground level to reach the lowest branches. When securing the tree proper bands should be used, as these can be adjusted as the tree grows. They can be purchased from most gardening shops. They are rather like belts with buckles which can be released as the tree grows. Wire should be avoided, since it will damage the tree. The number of bands needed varies according to the size of the tree. A large tree needs three, a smaller specimen two. In the case of larger trees there should be bands at the extremities of the stem, i.e. near the ground and near the top branches and a third in the middle. Having prepared the hole ready for planting, check the tree. If it is container grown this should be removed carefully and the tree then placed, complete with its ball of soil, into the hole. In the case of a non-container tree, care should be taken to ensure that any roots which show signs of damage are carefully removed; these should be cut with a sharp knife. Once in the hole, fine top soil should be shovelled in until the roots are covered. The rest of the soil is placed in, a few spade-fulls at a time. To ensure that the tree is firmly planted, the soil should be trampled on. The soil at the top should be level with the soil mark on the tree stem.

Care after planting

Even the smallest tree consumes quite large quantities of water. During the summer when drought

PROPAGATING TREES

Trees can either be propagated from seeds/nuts/berries or from cuttings. The list below shows the correct method, when the material should be gathered and when the cuttings, seeds/berries/nuts should be planted. The following table should help you collect your fruits to propagate your own trees. You will see that with a number of species, e.g. the horse chestnut, planting can take place immediately or in the following spring.

SPECIES	WHEN TO COLLECT	WHAT TO COLLECT	WHEN TO PROPAGATE
Alder	Autumn	Cones (remove seeds)	Following spring
Ash	July/August	Winged seeds	Sow as collected or following spring
Aspen	January to March	Cuttings	As collected
Beech	October	Nuts	Sow now or next spring
Birch	August/September	Seeds - collect twig/dry	Following spring
Blackthorn	August/September	Berries - dry	Sow in second spring
Black Poplar	January to March	Cuttings	As collected
Sweet (Spanish) Chestnut	October	Nuts	As collected or following spring
Elder	August/September	Berries - dry	Sow in second spring
Elm	May/June	Seeds	When collected or following spring
Field maple	October	Winged seeds	As collected or following year
Gean	August	Berries	Sow second spring
Guelder rose	August/September	Berries - dry	Following spring
Hawthorn	August/September	Berries - dry	Sow second spring
Hazel	August/September	Nuts	Following spring
Holly	March/April or Nov/Dec	Berries	As soon as collected or following spring
Horse chestnut	October	Nuts	As collected or following spring
Lime	September	Fruits	Sow second spring
Oak	October	Acorns (nuts)	As collected or following spring
Rowan	August/September	Berries - dry	Following spring
Scots Pine	January	Cones – remove seeds	Following spring
Spindle	August/September	Fruits	Following spring
Sycamore	September/October	Winged seeds	As collected or following spring
Walnut	October	Nuts	As collected or following spring
Whitebeam	September	Berries	Following spring
White Poplar	January to March	Cuttings	As collected
Yew	September to November	Berries - remove seeds	Sow second spring

NB Where cuttings are taken these should be from two-year old growth.

conditions often exist it is necessary to ensure that watering is carried out in the cool of the evening, otherwise the sun tends to draw out the moisture, almost as soon as the water is poured into the soil. If the tree is planted in a grassed area, there should be a clear patch of soil around the stem. This is important while the tree is becoming established. Depending on the situation of your garden, you might find it necessary to protect your tree with a guard. If you live in an area which borders on open countryside, you might find that wild animals getting into the garden attack your seedlings and small trees. The bark of the tree is very important, acting as both a protector and an insulator. Should it become damaged the tree will suffer and perhaps die.

Care should be taken to check the tree regularly. Twigs in particular, and occasionally branches, do get damaged, and a sharp knife or pruning shears should be used to remove the offending material. The tree should be pruned regularly for the first few years.

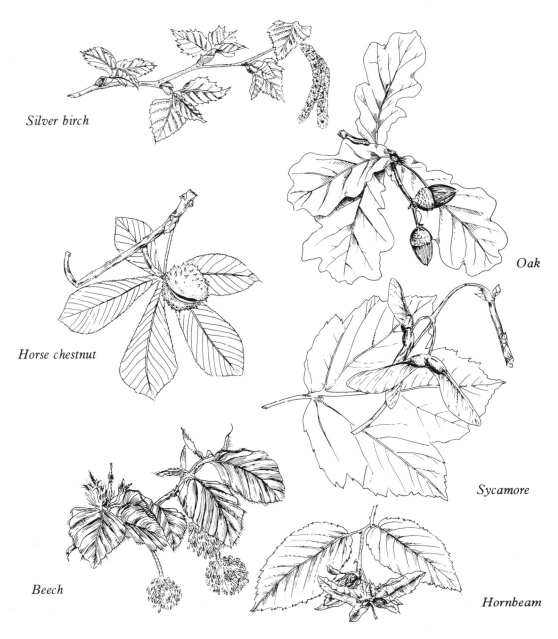

Silver birch

Oak

Horse chestnut

Sycamore

Beech

Hornbeam

Log Homes For Wildlife

In older methods of managing forests there would always have been good supplies of timber of all ages and at all stages of development, from the newly planted tree to the tree stumps left to decay. Unfortunately for wildlife, in many woodlands this is no longer so. There are some species of insects, like the large, harmless, but nevertheless ferocious-looking stag beetle, which must have rotten wood. The female will lay her eggs in the decaying timber, which acts as an important source of food for the grubs once they have hatched. Less rotting wood – less beetles; no rotting wood – no beetles. Then there are wasps which also use them, and indeed the rotting log will harbour a whole world of smaller wildlife. The digger wasps are quite partial to a log, where they can lay their eggs, and nearby they will look out for caterpillars and flies, which they can catch to provide a source of nourishment for their hungry growing offspring.

Similar to the wasp in appearance, but having only one pair of wings, is the hoverfly. Surviving under the guise of the wasp, a form of 'evasion' which is known as mimicry, helps to protect this creature from its enemies. The females of some species of this insect lay their eggs in rotting logs. But perhaps you wonder why you should encourage these insects to your garden. There are some species which are as effective as ladybirds, in devouring large numbers of aphids – greenflies.

So while you are out in the countryside, and without doing any damage, pick up a few rotten logs and take them back to 'plant' in your garden. Once established they will prove a valuable home for many different species of invertebrates and the chances are that some of the species which your logs encourage will be useful in your garden. You will want to take a peep to see what new inhabitants have taken advantage of your hospitality. Do so discreetly, and you won't do much harm. But also beware, because most of the animals which live there will be fast movers, and the chances are that you will hardly see them, before they have made their getaway!

A home for many animals

Both slugs and snails will take to the habitat in the damp ground beneath the log. Although some species of slugs are a nuisance in the garden, there is one, the great grey slug, which does no damage because it lives on a diet of decaying leaves and fungi. Leave it in your garden and one day you might thrill to its acrobatics as it performs its courtship ritual. Having secreted a thick length of slime, it hangs suspended from a tree or other suitable object and performs a ritualistic display.

Snails also need a moist atmosphere, and the log usually satisfies their requirements. They tend to alternate between periods of activity and prolonged periods of rest. These sleeping sessions are brought on by spells of dry weather, and are not the same as the winter hibernation. For their once-yearly deep sleep they will secrete a number of layers of slime over the entrance to the shell, in an effort to keep out the penetrating cold. In extreme spells of low temperatures even this precaution does not save them from the cold, and many will often die.

The large garden snail, a frequent inhabitant of walls

The centipede and the millipede are quite often confused, but with further investigation the problem as to which is which can be solved. Millipedes have rounded bodies: centipedes are more flattened. The former are carnivorous by nature, and in order to catch their food they have to move quickly. On the other hand, the millipedes are herbivores, and so they move at a much more leisurely pace, since the plants and decaying material on which they feed are there for the picking. The poison fangs of the centipede, which will quickly inject their lethal liquid into the unfortunate prey, ensure that there is no getaway. You might have time to look more closely at the millipede as it makes its leisurely way across your garden. You will see that there are two pairs of legs to each segment: in the centipede there is only one.

These small animals would quickly become desiccated if they did not have the moist atmosphere which exists under a log, and under stones as well. The woodlouse is another species which falls into the same group. Although earthworms are to be found in other parts of your garden, they will tend to be near the surface under the log, again because they can benefit from the dampness which is generally a feature.

If your log still has some bark left on it, you are providing yet another habitat for those species which can cope with slightly drier conditions. Here earwigs usually abound: other species of woodlice are also likely to view such a situation favourably.

Over a period of time the log will begin to deteriorate, much of this brought about by unseen plants, as they act on the wood. Once the log has begun to soften it will be attractive to those solitary wasps and bees which need such a place in which to lay their eggs (see page 89).

The Importance Of Hedges

By far the greatest asset to any garden is the hedge; it is a refuge for birds, and a home for several species of animal. To some extent its value is in the way in which it is managed, and in the type of hedging material which is used. Of all the places in the garden it is the only one which offers wildlife a screen, a place to hide, a place which gives them a certain amount of privacy and seclusion from the ever-present threat of man. The hedge also has other values, which we perhaps don't usually think about. Our weather, as well as being totally unpredictable, shows great extremes. A hedge around a garden helps in various ways. It acts as a windbreak and so encourages birds to use the garden, and even seek refuge there when open country round about might prove too hostile. It provides cover and shelter, not only during the nesting season but at other times of the year, and can be a valuable roosting site.

In the countryside in general the mixed hedges are those which often support the best and most varied bird population. Like trees, the hedge is something which takes a while to 'mature', but it will become established more easily and more quickly if various actions are taken, as we will explain shortly.

If your garden is large enough, you may feel that you have room for other hedges as well as boundary hedges. If, as was suggested in Chapter 2, you turn your garden into a bird sanctuary or have a rough area, you might find it is a good idea to plant a hedge here, to separate your wild areas from a more formal part of your estate.

Evergreen hedges

You will obviously want to know what sort of hedge to plant. Mixed hedges are very useful, but often the length of garden available is too short. There are many advertisements in the gardening press and the gardening sections of the popular press for quick-growing conifers. If you want a quick-growing hedge then one of these species, like Lawson's Cypress, will fit the bill. Because of its rapid growth it needs very severe pruning since, unlike hawthorn and blackthorn, it is a tree rather than a shrub. Nevertheless, it does quickly form a thick, tall hedge. This variety is quite cheap to purchase, but beware of 'bargain' offers, because such trees may either be less than the best or not have been hardened off. Although more expensive, there is a species which grows more quickly than Lawson's Cypress: this is Jackman's Green Hedger. You will have probems with these particular trees if you live close to the sea, where your garden is affected by salt spray. As an alternative, there is a more hardy species. Known as *Cupressocyparis leylandii*, it is best obtained in pots or containers. There are various names, but one of the most popular, and one which is recommended is a species known as *Castlewellan*. In a good soil this species will grow at the rate of between 40 and 50 cm (16 and 20 in) per year. Once established it forms a very dense, wind resistant hedge. For a really good barrier it is a useful idea to plant the trees in a double line, the plants spaced at intervals of about 60 cm (2 ft). Within a short time the main shoot will show quick growth. You should allow this to continue until you have the desired height: at this point it is time to trim it.

Another fast-growing conifer is *Thuja plicata*,

and although it does grow quickly it needs attention from the shears only in the summer months. Some species are susceptible to too severe clipping, but not this one, which makes it a useful variety. It can be cut back, but will still produce plenty of growth. Birds are quite attracted to it, and find it very acceptable for nesting sites. It is best to select younger trees of between 50 and 60 cm (20 to 22 in) in height. These should take well, provided they are planted in May. There should be about 90 cm (3 ft) between one tree and the next. They will reach a maximum height of about 2 m (7 ft) and, after planting, growth should be allowed to continue until the trees reach the height which you think is suitable for your requirements.

A species which has always been popular as a hedging plant in gardens, although it has often been kept too short to be of any value to birds when selecting their nesting sites, is *Buxus sempervirens*, or box. It has a preference for chalky soils, like those of the Downs, and here it grows well. Indeed, one of the few places in Britain where this species grows wild is the chalky Box Hill in Surrey. Here the shrub thrives on the exposed hillside, and the air is filled with its pungent scent.

It is particularly suited to the western parts of our shores, where a combination of climate and soil are advantageous to its growth. Here it can reach a maximum height of 1.5 m (6 ft). Not only does it grow upwards, but unless checked and trimmed regularly it will grow untidily in width. It is best planted in early spring, late in March if the ground isn't frozen, and strong plants should be purchased. Individuals should be placed about 60 cm (2 ft) apart. Because of its very thick foliage and a good branching system it is often favoured by the smaller birds. Not a difficult hedging species to maintain, it will only need clipping about once a year. Of course, if you really want to be ambitious, you can try your hand at topiary instead of simple trimming, as box is the shrub most frequently used for this art!

Deciduous varieties

Of course there are advantages to evergreen hedges. You have green throughout the year, and your garden is not subjected to the 'problem' of leaf fall in autumn, although you will miss the leaves for your compost bin. There may be some problems though, and the usual ground-layer flora associated with mixed hedges may either be absent or very restricted, whereas with a mixed deciduous hedge, before the leaves appear there may be a considerable growth of plants, providing a habitat that supports many small animals and insects. The presence of the insects will attract such creatures as the hedgehog and the wood mouse, enticed by the mixed diet they afford.

As an alternative to the conifers, perhaps nothing looks better in the garden than a hedge of copper beech, especially in autumn. Or you could use 'green' beech instead. There are advantages and one of the most important is that because this species is in leaf quite early in the year it is likely to attract some of the first nesting birds. In addition, you will have plenty of leaves in autumn, when, by the way, this species needs trimming back. Its maximum height of around 2.5 m (8 ft) provides a good screen. As with conifers the main growth points should be allowed to carry on to the desired height before cutting back is attempted. Although the beech is a deciduous species, when used as a hedge it will retain much of its foliage from one year until the following spring, when the new growth starts to appear. It takes time to establish a good beech hedge when compared with some of the fast growing evergreens, but if you have the time and patience you will be well rewarded.

When the enclosure Acts and Awards of the eighteenth and nineteenth centuries demanded the fencing in of fields, many landowners turned to hawthorn and, to a lesser extent, to blackthorn. The reason for their choice was that both are relatively fast-growing shrubs. But another attribute is that, although deciduous and devoid of leaves from autumn through to spring, well trimmed and maintained these hedges provide a very good supply of haws or sloe berries in the autumn, which is much appreciated by many different birds. As a garden-hedging species, both have similar properties. Haws in particular will attract birds in the colder months of the year. Growth is good, with a maximum of about 35 cm (15 in) annually. A maximum height of 1.5 m (6 ft) can be expected, and this will probably be achieved in five or six years. Soils, as well as management, do affect growth rates, and it is a good idea to have a soil test, so that you can get advice on management from your local nurseryman or garden centre.

Again, this species is very useful because it is one of the earliest hedgerow shrubs to come into leaf, and is therefore much sought after, particularly by the earlier nest-builders, although later ones will also use it. Its growth form provides thick, almost impenetrable foliage which provides a good selection of sites for nest-building. Among the first of the new season's builders will be dunnocks, willow warblers, thrushes, chaffinches, pied wagtails and blackbirds.

Hawthorn is not only valuable as a species which will provide cover for birds; there are lots of other animals, mainly invertebrates, which feed on the foliage. To provide a good dense hedge, plant the shrubs about 30 cm (12 in) apart. You will be pleased with your hawthorn hedge for most of the year. The white flowers warrant the constant attention of bees and other winged insects, then the autumn fruits complete the seasonal offerings.

The holly hedge

Holly is another species which comes high on the list for providing a good hedge, with some naturalists suggesting that it makes the 'best' hedge of all. This is probably true as far as birds are concerned, providing a hedge with a dense growth, which other species, and particularly the unpredictable one – man – have great difficulty in penetrating. Although it is necessary to have both male and female trees in order to have a good crop of berries, there is another feature which many people will take into account. The plants are rather costly, and this is one of the reasons why so few gardens have holly hedges. Added to this holly is generally slow-growing, and these are the two main reasons for not planting it in your garden. If you can afford it, do think seriously about holly for your hedges. There are a number of cultivated species, as opposed to the natural variety, and these tend to grow more quickly, although the fastest will average no more than 15 cm (6 in) in any year. Such a growth rate will take

place only on good soil, where the ground is well manured, and to have a hedge 1.5 m (6 ft) high will take at least twelve years. There is no reason why the holly should not be used in a mixed hedge, and hawthorn is useful here.

The right hedge for your soil

Soil conditions have already been mentioned, as providing a clue to the type of species which you could grow in your garden. Box, beech and yew thrive well in chalky soil conditions. Don't forget that yew is poisonous, and is best avoided where children are likely to use the garden. One species, *Taxus baccata* will form a very dense barrier. The foliage, which is dark, almost black, acts as a very good backcloth for other brighter hedgerow plants, which you might want to include in front of your screen of yew. If you have clay soils you will find that hawthorn does particularly well. Hornbeam is also useful, and both these are species which can tolerate the sticky conditions. The catkin-bearing hornbeam makes an attractive hedge that will withstand close clipping, making a leafy shelter for small animals. Similar to beech in appearance, it has the same characteristic of keeping its autumnal leaves throughout the winter.

One of the standbys for any garden is the privet. Now that there are a number of varieties available, there is no reason why there shouldn't be one to suit most people's tastes. Unlike many of the species which we have mentioned so far, privet needs attention throughout the summer, so there is a possibility of disturbing some species later in the year when clipping the hedge. Although its growth does not provide the same sort of dense cover as, say, the hawthorn, this can be overcome by planting a double hedge, consisting of two rows, particularly if individuals are planted 35 cm (15 in) apart, using a staggered arrangement with about 45 cm (18 in) left between the two rows. There are a number of varieties on the market. The most popular is *Ligustrum ovalifolium*. Although an evergreen species, in a very exposed situation it is likely to lose its leaves.

Wild shrubs for your hedge

We have already touched on the blackthorn. Known also as sloe, from the fruit, it is the

PLANTING GUIDE TO SPECIES SUITABLE FOR HEDGING

SPECIES	PLANTING DISTANCES		TIME OF YEAR TO	HEIGHT	
	Cm	Ins	Prune (P) Trim (T)	M	Ft
Berberis	38-46	15-18	February (T)	1.8	6
Cupressocyparis leylandii	76-92	30-36	July (T)	3.7	12
Forsythia	92	36	April (T)	1	3
Holly (*Ilex* spp)	61	24	August (T)	3	10
Lavender (*Lavandula* spp)	23-30	9-12	March - late (T)	0.6	2
Laurel (*Prunus laurocerasus*)	46-61	18-24	July - late (T)	5	8
Lawson Cypress	30-61	12-24	April - mid (T)	3.7	12
Privet (*Ligustrum* spp)	30-38	12-15	As necessary (T)	3	10
Pyracantha (Firethorn)	61-92	24-26	July - late (T)	3	10
Rosa rubiginosa	76-92	30-36	Late winter (P)	1.2	4
Roses (Floribundas)	46	18	February - late (P)	1.2	4
Spire	46-61	18-24	After flowering (T)	1.5	5
Thuja plicata	46-61	18-24	Summer - late (T)	3.7	12
Beech (*Fagus sylvatica*)	38	15	October (T)	2.4	8
Box (*Buxus*)	24	61	Autumn (T)	1.5	5
Escallonia macrantha	36	92	March - late (T)	1.8	6
Fuschia	92	36	October (T)	1.2-1.5	4-5
Honeysuckle (*Lonicera nitida*)	38	38	September/October (T)	1.5	5
Hornbeam (*Carpinus betulus*)	45	18	October (T)	1.8	6
Musk roses	121	48	March/April (P)	1.8	6
Blackthorn (*Prunus spinosa*)	91	36	October (T)	2.4	8
Pyrus communis	91	36	September/October (T)	3	10
Dog rose (*Rosa canina*)	121	48	March/April (P)	1.5	5
Buckthorn	91	36	September/October (T)	1.8	6
Whitethorn	30	12	September/October (T)	1.8	6
Hawthorn (*Crataegus* spp)	30	12	October (T)	2.4	8

'ancestor' of our modern-day plums. Clusters of white sprays appear in spring before the leaves are out. It is naturally attractive to those insects which are out early, like the bees and, sometimes, wasps. There are a number of varieties, the common, wild planted species is usually *Prunus spinosa*. As its common name suggests, the wood is black and has many thorns. Birds find its dense cover useful for building their nests and so, around a garden, it will attract many species. You could also collect the fruits in autumn, to make wine or, as many country folk prefer, sloe gin. Although there are some cultivated forms, it is *P. spinosa*, the wild variety, which has the richest source of nectar. It grows quite vigorously and unless trimmed back in autumn it can become unmanageable.

Few people think of using the wild shrubs for their hedges, and although hawthorn might feature in garden hedges, blackthorn is not seen as often. There are other species too which can be used, and which, if tended carefully, and not allowed to get out of hand, will add a new interest to the garden. The wild roses should be considered, although, as you will realise, they tip-root easily and so spread very quickly. However, there is perhaps nothing more pleasant than the dog rose, *Rosa canina*, with its large pink flowers. Autumn comes and the flowers, long since faded, give rise to the familiar hips, which I am told were collected, particularly during the War, for their valuable vitamin C content. But of course the hips are best left to the birds, and to some extent to the mice, which seek them out in autumn, as a rich source of food. If you have plenty of garden and can leave a rough area for the wild rose to colonize it will almost certainly soon become a mecca for a variety of bird species.

You might like to try the sweet briar, *Rosa rubuginosa*, as an alternative. Like the dog rose its flowers, single and pink, which appear in summer, are an attractive feature. Species roses which are grown today in the garden as shrubs have been derived from wild stock, originally found in various parts of the world. Such roses can be traced back to the very earliest gardens of which we have any record. Take *R. gallica*, more popularly known as the Red Rose of Lancaster: there is proof that this species was one of those which the Ancient Greeks cultivated. In mentioning this species we would be in trouble if we didn't refer to *R. alba*, the White Rose of York. Although we do not know for certain, it is possible that this species was brought to the British Isles by our Roman ancestors. A distinct disadvantage in growing these wild roses is that the flowering period is quickly over. Two species recommended are *R. rubrifolia* and *R. moyessi*, because the latter has very large bottle-shaped hips and *R. rubrifolia* is generally grown because of its attractive foliage.

Ladybird devouring aphids on a garden rose

Cultivated roses

Apart from the wild roses, you might also want to plant some cultivated varieties, and there are a number which will fit the bill. The advantages are obvious in that they all provide good cover for those birds which start to nest later in the year, and in autumn they should provide a good crop of hips. Not only will the blooms add a very good splash of colour in summer through to autumn, but most of the roses will attract insects out in search of nectar and fill the garden with their scent. It is as well to remember that unlike the hawthorns and the evergreens, the roses will need some form of support at least in the initial stages. Since you want a hedge which will reach a height of about 1.5 – 1.8 m (5-6 ft), any trellis or stakes which you use should be at least this height. You've probably realized that a rose hedge will usually achieve a maximum of around this height, although there are some species which grow to

137

2-2.4 m (7-8 ft). Another factor with rose hedging, as with holly, is that the cost might prove prohibitive to many people. The rose hedge needs very little attention beyond the annual removal of dead wood and there is such a large number of species to choose from, that it would be wise to discuss your choices with your local garden centre. Amongst the varieties which present a good display of blooms are Penelope, Nevada, Heidelberg, Kassel, Chinatown and Charles de Mills.

Other fruit-bearing hedges
We could turn our attention to the barberry for a hedging plant. There are many different species which are evergreen shrubs, producing a good crop of berries in the autumn. The most popular are *B. rubrostilla*, *B. thunbergii*, *B. stenophyll* and *B. darwinii*. *B. rubrostilla* has a maximum height of 1.25 m (4 ft), with a spread varying from 1.8 m to 2.4 m (6-8 ft). It is particularly attractive in the garden, because it produces a mass of brilliant coral-red berries and exhibits a rich ruby leaf colour in autumn. *B. thunbergii* reaches a maximum height of 1.8 m (6 ft), and its greatest spread is 2.4 m (8 ft). In spring the shrub is characterized by its beautiful yellow flowers which give rise to brightly coloured red berries in the autumn. *B. darwinii* will reach a maximum height of between 2.4 and 3 m (8 and 10 ft), and its spread is usually about the same. The orange-yellow blossoms appear in May and June. The berries of this variety are purple-blue. The fourth species, *B. stenophylla* reaches the greatest height, with 3 m (10 ft) being a maximum, and it also has the widest spread at 3.7 m (12 ft). The golden flowers appearing in April are particularly pleasing to the eye. In autumn the berries will be blue, although they have a visible white sheen. Berberis can tolerate a wide range of climatic conditions, and a situation in full sun to partial shade. Like the rose, as far as soils are concerned, they are not particularly fussy and grow well on almost any type, including chalk. It is easy to propagate the plant by taking hardwood cuttings in the autumn.

A hedge by the sea
Those people who live along the coast may find it difficult with certain hedging species, as they are susceptible to the salt-laden air, and so do not do well. If you have been to some parts of the coast you will probably have been confronted at some time by a shrub with rather sharp spines. If you have been in autumn you will have seen yellow berries being taken by flocks of migrating birds. The shrub is the sea buckthorn, *Hippophae rhamnoides*. Within a short time its vigorous growth, aided by good sucker development will provide a dense hedge. It has a maximum height of about 1.8 m (6 ft), and needs virtually no attention, at least in the early stages. Some pruning may be necessary later, together with the removal of the sucker growth, otherwise it will take over the garden.

Spindle tree (*Euonymus japonica*) can be used and will form a useful hedge. 'Spindle *tree*' is something of a misnomer, for it is better described as a shrub. It is tolerant of salt spray as well as pollution. It reaches a maximum height of about 6 m (20 ft), although as a hedge species this is usually reduced to 3 m (10 ft). When planted at a distance of about 60 cm (24 in) it will grow to offer a substantial hedge, which will prove of value as a nesting site for many different species of birds.

Habitats In Walls

Walls can become an important habitat for wildlife, and there are ways of increasing their potential. The materials used for building the wall vary from one part of the country to another. The most common material is brick, particularly in modern walls. In some places local stone predominates, and if you go to some areas, like the Norfolk coast, you will see that large pebbles and flints from the beach have been used. These walls are valuable to wildlife because of the irregular spaces between the stones. Slate, limestone, ironstone, granite, flint and sandstone, all these feature in walls to a greater or lesser extent. If you are fortunate enough to have a wall made of limestone which is not newly constructed, you will probably already have realized how rich it is as a habitat for wildlife.

The importance of aspect
The aspect of the wall is of great importance to the animals and plants which will colonise it. Where walls are aligned in such a way that they attract

the attention of the sun for large parts of the day, the wildlife which they contain is usually negligible. Plants and animals are unable to tolerate not only the intensity of the sun, which heats up the wall, but also the drying effect which this has. Contrast such a south-facing wall with one which has a north-facing aspect and, as you will see, the change is very great. Here the wall remains cool and shaded, and where it is well established, will have a wide variety of plant species, not to mention the animals which take refuge either within the wall, or among the plants.

The maturing wall

We have mentioned that it is the old walls which are obviously richer, particularly in terms of plant life, than the newer ones. Over a period of time, the elements will have taken their toll or done their work, and this weathering is very important, because it will have been effective in breaking down some of the mortar and cement between the building blocks, releasing nutrients necessary for the well-being and development of wall plants. Coupled with this is the fact that over a period of time soil blown about by the wind will collect in nooks and crannies and provide a supply of food for the plants.

The first colonizers

Which are the first plants to colonize what appears to be a bleak and barren substrate, lashed by rain and wind, and perhaps scorched by the summer sun? The first species to arrive will be the lichens. Indeed, it is generally only the lichens which manage to survive the uncompromising conditions which a south-facing wall presents. Initially the form of these simple plants is so small that they cannot be detected by the human eye. Gradually they spread out, often forming circular patches, although this shape is by no means unique. Their growth rate is extremely slow, but once they are established they survive for a very long time. It is possible to gain some idea of the age of a wall by looking at the size of the lichen patches. Similarly, by looking at gravestones, which have dates on them, it is possible to work out their rate of growth. Lichens reproduce by means of spores, and it is because these are very small and light that they are quickly carried to new habitats. Various types of lichen are distinguished by their colour,

and you may find grey, orange or yellow forms. All glory in Latin names; none has a common English one!

The arrival of moss

Where some soil appears on a wall, the spores of moss may also settle and grow. If you look at various walls you will see that there are several species of moss which grow, although by far the most common is *Tortula muralis*, known as wall screw moss, whose bright green form enlivens many walls.

Next the wall will support several flowering plants, once the habitat is suitable for them to gain a foothold. You may find the wall pennywort and the wall barley, although they may not be exclusive to the wall. Before these species can become established, a reasonable supply of soil has to accumulate.

There are times when no soil is visible, but plants have become established. Where either weathering or poor workmanship has provided a supply of loose mortar, these plants will have something in which to sink their roots. One species which has taken advantage of many a niche which the wall offers is the rosebay willow herb. But there is a wide range of plants which will grow on the wall provided that the conditions are favourable, and I remember watching with interest as an elder tree grew out from between the blocks of stone in the village church tower where I grew up as a lad. It obviously found plenty of nourishment, because in a relatively short time it reached shrub size.

Light seeds

Nature has provided various plants with suitable adaptations so that the species will continue. Many species have very light seeds, and as the wind carries these along a lot unexpectedly come up against a solid barrier, the wall, and from time to time the odd seed falls, where there is a suitable place in which to survive. Where a wall presents a shady aspect, other plants including ferns, will grow as well as the flowering species. The maidenhair spleenwort will grow where conditions are not too adverse, and will flourish as will the wall rue. It stands to reason that species like these are adapted for the demanding conditions which are found on and in the wall.

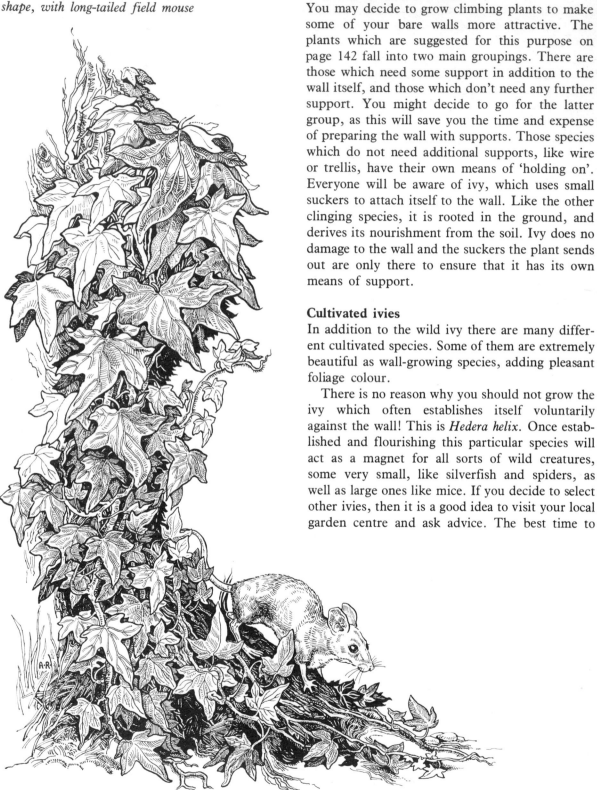

Roots and stem of ivy showing the five-pointed leaf shape, with long-tailed field mouse

Climbing species for walls

You may decide to grow climbing plants to make some of your bare walls more attractive. The plants which are suggested for this purpose on page 142 fall into two main groupings. There are those which need some support in addition to the wall itself, and those which don't need any further support. You might decide to go for the latter group, as this will save you the time and expense of preparing the wall with supports. Those species which do not need additional supports, like wire or trellis, have their own means of 'holding on'. Everyone will be aware of ivy, which uses small suckers to attach itself to the wall. Like the other clinging species, it is rooted in the ground, and derives its nourishment from the soil. Ivy does no damage to the wall and the suckers the plant sends out are only there to ensure that it has its own means of support.

Cultivated ivies

In addition to the wild ivy there are many different cultivated species. Some of them are extremely beautiful as wall-growing species, adding pleasant foliage colour.

There is no reason why you should not grow the ivy which often establishes itself voluntarily against the wall! This is *Hedera helix*. Once established and flourishing this particular species will act as a magnet for all sorts of wild creatures, some very small, like silverfish and spiders, as well as large ones like mice. If you decide to select other ivies, then it is a good idea to visit your local garden centre and ask advice. The best time to plant ivy is in the spring, so that it will become well established in the ensuing months. Unless you buy container- or pot-grown species, you may find that they do not do too well if the rooting system is upset, as they don't take kindly to being moved. If you dig in some rotted manure you will provide the plants with nourishment, and give them a good start. The foliage of ivies varies from the dark green of species like *Hedera helix, var. caenwoodiana*, to the variegated leaf of *H. colchica, var. dentata variegata*. In between these extremes are many other subtle hues and shades and you might decide that a mixture would give added variety to your wall. The part of the country in which you live will, to some extent, determine the species which you buy, since some are particularly hardy, whereas others are less tolerant of severe conditions and might suffer in extremes of temperature.

The world of the ivy

If you already have ivy on the wall of your house you will perhaps already have discovered in it a hidden world of wildlife. Because the ivy is ever-green, and because it flowers and fruits much later than most other plants, it is particularly valuable in providing shelter and to some extent, food, all the year round. Although the branches higher up the wall will offer some protection, it is around the base where the ivy is at its thickest, that wildlife will seek shelter, not only during the spring and summer, but in the autumn and winter as well. Small tortoiseshell butterflies may settle behind the leaves to spend a quiet winter. There will be many other animal species which will come on the prowl for food, and foraging ants and millipedes may be overlooked. In very thick growth mice may find a suitable place to rest, and birds are also likely to seek shelter.

Nesting sites

The value of ivy and other evergreen creepers is that you will, hopefully, have a ready-made nesting site for your early spring nest-building birds. Among those which are the first to build are the blackbirds, species of thrush, the chaffinch and the chiff-chaff. As most other plant species are fading, the ivy will be bursting with flower in October, especially where there are shaded, rather than exposed, walls. Late insects, particularly if

The top of mature ivy, showing berries in spring, and some heads which have lost their berries. There is also an indication that the shape of the leaf changes on different parts of the plant

the weather is fine and warm, will take eagerly of the nectar which the ivy flowers offer. The familiar blue and black berries, which appear early in the following year, will provide a source of food and nourishment for birds when other natural foods may not be easy to come by, especially in hard winters.

If you have a hedge in your garden which comes into leaf early, and so becomes the target for the early nesters, you might want to include some creepers because their leaves appear later and will provide a nesting site for those species which build their nests then, like the migratory birds which do not arrive here until later on. You might want to consider the value of encouraging garden warblers and greenfinches to nest in your garden, and perhaps, with a species which comes into leaf later, you might even tempt a flycatcher to suss out your territory.

PLANTS FOR AND ON WALLS

For growing in spaces between the building blocks

Ferns

Adiantum capillos veneris   *Asplenium adiantum nigrum*

Asplenium lanceolatum   *Asplenium ruta muraria*

Asplenium trichomanes   *Phyllitis scolopendrium*

Polypodium vulgare

Other plants

House leek (*Sempervivum tectorum*)

Ivy-leaved toadflax (*Cymbalaris muralis*)

Rue-leaved saxifrage (*Saxifrage tridactylites*)

Stonecrop/Wallpepper (*Sedum acre*)

Wall pennywort (*Umbilicus rupestris*)

Wall rocket (*Diplotaxus arvenis*)

Wallflower (*Cheiranthus cheiri*)

Male fern (*Dryopteris felix-mas*)

Plants for growing against the wall

Camellia spp   Trumpet creeper

        (*Campis grandiflora*)

Clematis spp   Cotoneaster spp

Broom (*Cytisus battandieri*)   *Eccremoncarpus*

Garrya elliptica   Ivy (*Hedera spp*)

Caelastrus scandens   Hydrangea spp

Wisteria   Pyracantha spp

Lathyrus   Jasmine (*Jasminium* spp)

Russian vine (*Polygonum baldschuanicum*)

Rose spp   *Parthenocissus* spp

Mosses found on walls

Rough-stalked feather moss (*Brachythecium rutabulum*)

Cushion moss (*Grimmia apocarpa*)

Beard moss (*Barbula unguiculata*)

Silky feather moss (*Camptothecium sericeum*)

Wall screw moss (*Tortula muralis*)

Garden thistle (Centaurea hypdeuca)

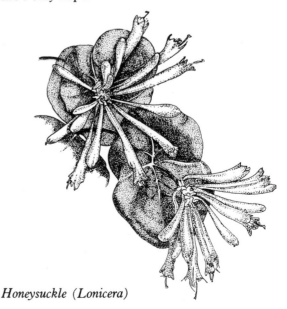

Michaelmas daisy (Aster amellus)

We hope that in this chapter we will have given further ideas about planning your garden to encourage wildlife. But if you have only a small plot and feel that trees and hedges are not for you, do not despair: even if your garden is only a back yard there is a lot you can do. If you can only hang up a bag of scraps you will entice the birds in to feed, and you can always suspend hoppers and feeding trays from the walls for them. To attract the small mammals who have probably already paid a passing visit, grow creeping plants up your walls, and while you are waiting for them to establish themselves, remove one or two bricks and plant trailing lobelia, creeping jenny or even the perennial aubretia in the cavities after filling them with potting compost.

All plants will attract insects, and there must be room for containers for annuals or even small permanent shrubs. Even window boxes will be a magnet for bees and butterflies, and there is a large variety available now. These and hanging baskets can be full of colourful scented plants well into the autumn.

Finally, try making a mini-pond from an old sink as suggested on page 102, and you may be surprised at the wildlife it will attract.

However little you can do to help sustain wildlife on your territory, you will find it gives you an endless source of interest, and it may be the wildlife's only hope!

Primula

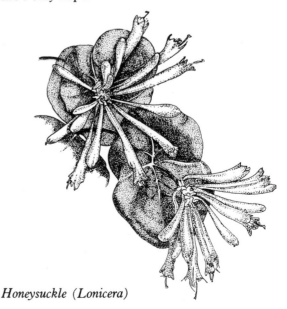

Honeysuckle (Lonicera)

Further Useful Information

Organisations and Societies

The Arboricultural Association, Brokerswood House, Brokerswood, Westbury, Wilts, BA13 4EH

Has various categories of membership, and publishes pamphlets, *etc*. about trees

British Naturalists Association, Hon. Membership Secretary, 23 Oak Hill Close, Woodford Green, Essex

Membership of the Society is open to individuals and to families. In addition there are also groups throughout the country. These arrange indoor meetings, excursions, etc. BNA produces *Country-Side*, which is the best general natural history magazine: it is published three times a year. Membership is worthwhile just for the journal.

British Trust for Entomology, 41 Queen's Gate, London SW7

Concerned with the study of entomology.

Conchological Society of Great Britain and Ireland, Mrs E. B. Rands, 51 Wychwood Avenue, Luton, Beds, LU2 7HT

The organisation organises regular meetings, and publishes many pamphlets and journals in connection with its interest in shells of all kinds.

Conservation Society Ltd, 12a Guildford Street, Chertsey, Surrey, KT16 9BQ

Concerned with conservation in many different forms.

Men of the Trees, Crawley Down, Crawley, Sussex

Produces the useful journal, *Trees*, and has various leaflets, *etc*., to help people understand trees, and increase the numbers in the countryside.

Trees for People, 71 Verulam Road, St Albans, Herts

Aims to encourage the increased planting of trees.

Periodicals and Magazines

Entomologists' Gazette, 353 Hanworth Road, Hampton, Middlesex

Entomologists' Record and Journal of Variation, E. H. Wild, 112 Foxearth Road, Selsdon, Croydon, Surrey

Books, Leaflets, etc

Ashbery, A., *Miniature Gardens*, David & Charles

Bristowe, W. S., *The World of Spiders*, Collins

Civic Trust, *Tree Leaflets*, from 17 Carlton House Terrace, SW1Y 5AW

Angel, H., *Wild Animals in the Garden*, Jarrold

Colvin, B. & Tyrwhitt, J., *Trees for Town and Country*, Lund Humphries

Colyer, O. N. & Hammond, C. O., *Flies of the British Isles*, Warne

CPRE, *Tree Sense*, from 4 Hobart Place, London SW1W 0HY

Edwards, G., *Wild and Old Garden Roses*, David & Charles

Ellis, E. A., *British Snails*, Oxford University Press

Ellis, E. A., *Wild Flowers of the Hedgerows*, Jarrold

Fitter, R., Fitter, A. and Blamey, M., *Wild Flowers of Britain and N W Europe*, Collins

Fitter, R. S. R., *Finding Wild Flowers*, Collins

Genders, R., *Covering a Wall*, Robert Hale

Genders, R., *The Wildflower Garden*, David & Charles

Genders, R., *Growing Old-Fashioned Flowers*, David & Charles

Gorer, R., *Trees and Shrubs*, *A Complete Guide*, David & Charles

Grounds, R., *Trees for Smaller Gardens*, Dent

Growing Native Trees and Shrubs to form a Wildlife Area, Farming and Wildlife Advisory Group, The Lodge, Sandy, Beds

Growing Trees and Shrubs from Seeds, Devon Trust for Nature Conservation

HMSO, *Know Your Broadleaves*, Forestry Commission

HMSO, *Know Your Conifers*, Forestry Commission

Hedges and Local History, Standing Conference on Local History, 26 Bedford Square, London WC1B 3HU

Hillier, H. G., *Manual of Trees and Shrubs*, Hillier & Sons, Winchester

How to Begin the Study of Harvest Spiders, BNA

How to Begin the Study of Crickets and Grasshoppers, BNA

How to Begin the Study of Entomology, BNA

How to Begin the Study of Spiders, BNA

Hyde, G., *Insects in Britain* (Books 1-4), Jarrold

Hyde, G. E., *Trees*, Town & Country Series, Almark

Institution of Municipal Engineers, *Tree Preservation and Planting*, A Protection of the Environment Monograph

Linssen, E. F., *Beetles of the British Isles*, Warne

Linssen E. F. & Newman, L. H., *Observer's Book of Common Insects and Spiders*, Warne
McClintock, D., *Wild Flowers*, Collins
Mitchell, A. F., *Common Trees*, Forestry Commission (HMSO)
Mitchell, A., *Trees of Britain and Northern Europe*, Collins
Nichols, D., Cooke, J. & Whiteley, D., *Oxford Book of Invertebrates*, Oxford University Press
Pollard, D., Hooper, M. & Moore, N., *Hedges* (New Naturalist), Collins
Proper Care of Trees, The Scottish Civic Trust, 24 George Street, Glasgow G2 1EF
Randall, R. E., *Trees in Britain: Broadleaves—1, 2, & 3*, Jarrold
Randall, R. E., *Trees in Britain: Bushes and Shrubs*, Jarrold
Riley, N. D. (Ed.), *Insects in Colour*, Blandford
Step, E., *Wayside and Woodland Trees*, Warne
Tree Planting and Wildlife, SPNC, 2 The Green, Nettleham, Lincoln
Tree Planting and Wildlife Conservation, Nature Conservancy Council
Wildlife Conservation and Dead Wood, Devon Trust for Nature Conservation
Wilson, Ron, *The Hedgerow Book*, David & Charles
Wootton, A., *Discovering Garden Insects and Other Invertebrates*, Shire
Wragge, Morley D., *Ants*, Collins

Equipment and Supplies

Austin, David, Roses, Albrighton, Wolverhampton:
Old fashioned shrub and climbing roses
Bees Ltd, Sealand, Chester:
Leylandii
Bressingham Gardens, Bressingham, Diss, Norfolk IP22 2AB:
Plants, dwarf shrubs, etc
Brown, D. T., & Co. Ltd, Station Road, Poulton le Fylde, Blackpool:
Untreated seeds
Buckingham Nurseries, Tingewicke Road, Buckingham:
Hedging shrubs
Cheal, J., & Sons Ltd, Stopham Road, Pulborough, Sussex:
Trees and shrubs
Christie, T. W., The Nursery, Forres, Morayshire, Scotland:

Trees
Civic Trees (Scotland) Ltd, The Gardens, Polton Lasswade, Midlothian:
Trees
Countrycraft, Robertsbridge, Sussex:
Rustic poles
Cowley Wood Conservation Centre, Cowley Wood, Parracombe, North Devon:
Wild plants, seeds, etc.
Cunningham, J., Verulam Nursery, Verulam Road, St Albans, Herts:
Cowslips
Cunningham, W. & Sons, Dewdrop Nursery, Heacham, Norfolk:
Shrubs, perennials, dwarf border plants
Dingley Nurseries, 12 Princess Mews, Belsize Village, London NW3:
Leylandii
EFG (Nurseries) Ltd, Fordham, Ely, Cambs, CB7 5LH:
Trees
Ferguson, W. M., 72 Campsie Road, Kildrun, Cumbernauld:
Trees
Forest Drums Ltd, P.O. Box 60, Southampton, SO9 7ED:
Plastic water butts
Hillier & Sons, Winchester, Hants:
Trees and shrubs
Hillside Nursery, Pointon, Sleaford, Lincs:
Pampas grass
Holtzhausen, 14 Hillcross Street, St Austell, Cornwall:
Wide range of plants and bulbs
Hydon Nurseries, Hydon Heath, Godalming, Surrey:
Trees and shrubs
Jackmans Nurseries Ltd, Woking, Surrey:
General
Kaye, Reginald, Waithman Nurseries, Silverdale, Lancs:
Herbaceous plants and some water plants
Knaphill Nurseries, Lower Knaphill, Woking, Surrey:
Trees and shrubs
Landscape Services & Supply Co, Conway Nursery & Garden Centre, Station Road, Lower Standon, Henlow, Beds, SG16 6NY:
Cupressus leylandii, green and gold, conifers, silver birch, *etc.*

Laxton & Bunyard Nurseries, Old Rydon Lane, Exeter, Devon:
   Trees and shrubs
MacDonald Bros, Bogton Nurseries, 67 Muirend Road, Glasgow:
   Trees
Newlands Conifer Nurseries, Myerscough Hall Drive, Bilsborrow, Garstang, Preston, Lancs:
   Conifers
Notcutt, R. C., Woodbridge, Suffolk:
   Trees, shrubs and other plants
PCM, 19-27 Kents Hill Road, Benfleet, Essex:
   Plastic covered terylene rope, plant supports
Panton, John, Coombe House, Exbridge, Dulverton, Somerset:
   Japanese honeysuckle and other plants
Perrie Hale Nursery, Honiton, Devon:
   Cupressus leylandii
Pickards Magnolia Gardens, Stodmarsh Road, Canterbury, Kent:
   Magnolias
Roger, R. V., Ltd, The Nurseries, Pickering, Yorks:
   General plants
St. Bridget Nurseries, Old Rydon Lane, Exeter, Devon:
   Trees
Scott, John & Co., The Royal Nurseries, Merriott, Somerset:
   Trees
Smith, James & Sons, Darley Dale, Derbyshire:
   Trees and shrubs
Southview Nurseries, Eversley Cross, Herts:
   Old fashioned granny's pinks
Stewarts (Ferndown) Nurseries Ltd, God's Blessing Lane, Broomshill, Wimborne, Dorset:
   Trees and shrubs
Sunningdale Nurseries, Windlesham, Surrey:
   Trees
Thompson & Morgan (Ipswich) Ltd, London Road, Ipswich, Suffolk, IP2 0BA:
   Seeds of wild flowers, trees, etc
Trenear, Peter, Chantreyland, Eversley Cross, Herts:
   Aubretia
Treseeders Nurseries (Truro) Ltd, The Nurseries, Truro, Cornwall:
   Trees and shrubs
Walmestone Nurseries Ltd, Wingham Well, Wingham, Canterbury, Kent:
   Trees
Wardington Nurseries, Banbury, Oxon:
   Leylandii, green and gold, in pots
Waterer, John & Sons & Crisp, The Floral Mile, Twyford, Berks
Weasdale Nurseries, Kirkby Stephen, Cumbria, CA17 4LX:
   Hardy trees, more than 600 species of shrubs and conifers for amenity and shelter purposes, screening and hedging
Welsh Tree Services, Abergavenny Road, Raglan, Gwent, NP5 2BH:
   Trees
Whines and Edgeler, Godmanstone, Dorchester, Dorset:
   Bamboo canes
Williamson (Hereford) H. Ltd, Wyevale Nurseries, Kings Acre, Hereford:
   Trees and shrubs

INDEX

Rowan 16, 27
Royal Society for the Protection of Birds
   bird feeder 22
   bird table 19

Slugs 132
Snails 132
Soil
   earthworm and 3
   nutrient cycle in 3
   types of 2
Solitary bees
   artificial homes for 88–89
Sorbus 16
Spindle 16, 138
Squirrels 18, 19, 56–57
Starlings
   problem of 16–17
Stickleback
   breeding habits of the 109
   food for the 108–109
   predators of the 108
   species of 108
Swallows
   nesting habits of 39
Sweet briar 137
Sweet rocket 79

Taxus baccata 16, 135
Thuja plicata 133–134
Toads, *see* Frogs
Tortula muralis 139
Trees
   care of 129–131
   choosing 124, 126–127
   decline in Britain 124
   feeding 129

   habitat, as 124, 125
   planting 124–128, 129
   site for 128
   time for 129
   propagating 128, 130
   seedlings, planting 128
   supporting 128
   value of 124

Upholsterer bees 90
Urine
   compost activator, as 6

Viburnum 15–16

Walls
   aspect, importance of 138–139
   climbing species 140
   habitats, as 138
   lichens 139
   maturing 139
   moss 139
   plants for and on 142
Wasps
   digger, *see* Digger wasps
   life-cycle of 92–93
   types of 92
Waterlilies 105
Weeds
   meaning of 2
   wild plants, as 6
Whirleygig beetles 110–111
Wild plants
   protection of 6

Yew 16, 135